Density Functional Theory - New Perspectives and Applications

Edited by Sajjad Haider, Adnan Haider and Salah Ud-Din Khan

Published in London, United Kingdom

Density Functional Theory - New Perspectives and Applications
http://dx.doi.org/10.5772/intechopen.104137
Edited by Sajjad Haider, Adnan Haider and Salah Ud-Din Khan

Contributors
Nurdiana Nordin, Subramanian Usha, Charles Kanakam Christopher, Masami Kusunoki, Abdelhamid El Kaaouachi, Brahim Ait Hammou, El Hassan Mounir, Hamza Mabchour, Driss Ennajih, Adil Echchelh, Said Dlimi, Abdellatif El Oujdi, Duduzile Zamavezi Nkomo, Bongani Ngobe, Maje Phasha, Ramogohlo Diale

First published in London, United Kingdom, 2024 by IntechOpen
IntechOpen is the global imprint of INTECHOPEN LIMITED, registered in England and Wales, registration number: 11086078, 5 Princes Gate Court, London, SW7 2QJ, United Kingdom

British Library Cataloguing-in-Publication Data
A catalogue record for this book is available from the British Library

Additional hard and PDF copies can be obtained from orders@intechopen.com

Density Functional Theory - New Perspectives and Applications
Edited by Sajjad Haider, Adnan Haider and Salah Ud-Din Khan
p. cm.
Print ISBN 978-1-83768-879-1
Online ISBN 978-1-83768-880-7
eBook (PDF) ISBN 978-1-83768-881-4

For EU product safety concerns:
IN TECH d.o.o., Prolaz Marije Krucifikse Kozulić 3, 51000 Rijeka, Croatia,
info@intechopen.com or visit our website at intechopen.com.

Meet the editors

Dr. Sajjad Haider is an associate professor in the Department of Chemical Engineering, King Saud University, Saudi Arabia. He received an MSc in 1999 and an MPhil in 2004 from the Institute of Chemical Sciences, University of Peshawar, KPK, Pakistan, and a Ph.D. in 2009 from the Department of Polymer Science and Engineering, Kyungpook National University, Taegu, South Korea. His research work focuses on the development of scaffolds for tissue regeneration, biopolymer composites, polymer hydrogels, fabrication of electrospun nanofibers and metal nanoparticles, and evaluation of their potential application in biomedicine and removal of hazardous materials from aqueous media. He has published more than 165 research articles in reputed journals and conferences. He has edited nine books and authored fifteen book chapters. He also holds two US patents, which are also registered in Europe and Saudi patent offices. He has been invited several times to speak at international workshops and conferences.

Dr. Adnan Haider is an associate professor in the Department of Biological Sciences, Division of Nanomedicine, National University of Medical Sciences (NUMS) Rawalpindi, Pakistan. He holds a Ph.D. and post-doctorate in Polymer Science from South Korea. His research work focuses on the development of scaffolds for tissue regeneration, biopolymer composites, polymer hydrogels, drug delivery systems, fabrication of electrospun nanofibers and metal nanoparticles, and evaluation of their potential application in biomedicine and removal of hazardous materials from aqueous media. Dr. Haider has 88 publications to his credit, including journal articles and book chapters. He is the editor of several books in his field of specialization. He is also the managing editor of Life and Science. He has been invited to speak at several international workshops and conferences.

Dr. Salah Ud-Din Khan is an associate professor at the Sustainable Energy Technologies (SET) Center, King Saud University (KSU), Saudi Arabia. Before joining KSU, he was an assistant professor, visiting scientist, researcher, and lecturer at various institutes and universities. His areas of research include clean energy sources, energy storage, radiological protection, and computational techniques for radiation shielding materials. Dr. Khan teaches and supervises postgraduate students in renewable energy technologies at SET Center. He has published more than 250 research articles in reputed journals and conferences. He has also authored three books and two book chapters. He has one US patent to his credit, which is also registered in Europe and Saudi patent offices. He has been invited as a speaker several times to international workshops and conferences.

Contents

Preface

Density functional theory (DFT) is a course that was registered as a "curriculum course" at the University of Munich in the 1990s, mainly covering topics such as time- and orbital-dependent functions. Over the years, DFT has proven to be a very useful theory in various chemical applications. This has prompted us to present to interested readers the most important breakthroughs and the current state of knowledge on the subject. Since the development of DFT as a quantum mechanical theory based on electron density, numerous efforts have been made to redefine important concepts already known in organic chemistry and make them useful for semiquantitative analysis. The book begins with the principles and theory from the ground states to the excited states of photoelectron emission. The detailed principles and theories are not limited to metal surfaces but also include nanoscale particles. Meanwhile, the DFT model (both time-dependent and time-independent) is widely used and has the advantage that it directly accounts for electron correlation in its formalism. In addition, DFT can be programmed for computer simulations to determine eigenvalues and eigenvectors. It provides enough information to determine the electronic structure and the total energy of the particles. An important discovery of Kohn-Sham density functional theory (KS-DFT), originally formulated half a century ago for Hamiltonians in isolated many-electron systems, was the driving force for materials science and led to a wide range of specialized codes for the prediction of molecular and crystalline properties. The first important improvement came from the discovery of the most important exchange-correlation energy functional, which led to the scheme provisionally referred to as the UHFD scheme. This scheme is the ultimate approximation beyond UHF and all other hybrid exchange-correlation energy functionals, suggesting that the search for a more powerful hybrid exchange-correlation energy functional may no longer be necessary to escape the chaotic state. The book also discusses the physicochemical properties of the drug ellipticine obtained by the DFT-B3LYP/6-311 G (d, p) method using the Gaussian computational package. HOMO (highest occupied molecular orbital), LUMO (lowest unoccupied molecular orbital), MEP (minimum energy paths) surface area, and chemical reactivity descriptors can be used computationally for practical applications.

Sajjad Haider, Ph.D.
Department of Chemical Engineering,
College of Engineering,
King Saud University,
Riyadh, Saudi Arabia

Adnan Haider
Department of Biological Sciences,
Nanomedicine Section,
PWD Campus,
National University of Medical Sciences,
Rawalpindi, Punjab, Pakistan

Salah Ud-Din Khan, Ph.D.
Sustainable Energy Technologies (SET) Center,
College of Engineering,
King Saud University,
Riyadh, Saudi Arabia

Section 1

DFT Theory and Practice

Chapter 1

The Use of DFT-Based *ab-initio* Technique to Determine the Stability Difference in B2 Ti-PGM Compounds

Ramogohlo Diale, Duduzile Nkomo, Bongani Ngobe and Maje Phasha

Abstract

In this chapter, the density functional theory (DFT) based first-principles approach is used to predict the underlying lattice properties associated with the phase transformation and stability of B2 phase in titanium-platinum group metal (Ti-PGM) compounds. This *ab- initio* technique provides a good platform to accurately explore phase stability variation between the successful Ti-PGM shape memory alloys (SMAs) ($Ti_{50}M_{50}$, M = Rh, Pd, Ir, Pt) and other B2 Ti-PGM compounds that do not show any shape memory effect (SME), such as $Ti_{50}Os_{50}$ and $Ti_{50}Ru_{50}$. The B2 TiFe, TiNi and TiAu have also been considered in this chapter in order to draw similarities and differences. Amongst the predicted results, the heat of formation was calculated to determine the thermodynamic stability, whereas the total densities of states were used to evaluate the electronic stability of these compounds. Insights on the mechanical stability of the B2 crystals were derived from the calculated elastic constants. Mechanical instability was revealed in some compounds, indicative of a possible phase transition responsible for the intrinsic shape memory character. Although an attempt to correlate this mechanical instability with imaginary frequencies established from the phonon dispersion curves is made, the correlation is not yet conclusive due to some discrepancies observed in TiNi.

Keywords: B2 Ti-PGM, shape memory alloys, DFT, thermodynamic stability, electronic stability, mechanical stability, lattice dynamic stability

1. Introduction

An interest in PGM-containing SMAs continues to rise due to their unique functional properties desirable for use in various high-temperature applications such as aerospace, automotive, power plants and chemical industries [1]. Ti-PGM intermetallic compounds such as TiPt [2, 3] and TiPd [4] have gained attention as promising candidates for development of high-temperature shape memory alloys (HTSMAs). This is so because they exhibit a martensitic transformation (MT) from cubic

CsCl-type B2 to orthorhombic AuCd-type B19 phase at temperatures higher (above 373 K) than that of a well-known commercial TiNi. TiPt has attracted significant interest in various industries due to its high MT from B2 to B19 at approximately 1273 K [2], whilst the MT for TiPd has been observed at 823 K [5, 6]. Similarly, for TiIr, its B2 phase undergoes MT at 2023 K to monoclinic phase [7], whereas the MT for TiRh binary system occurs at 1118 K from B2 to tetragonal $L1_0$ [8–10]. Similarly, the MT of TiAu from B2 to B19 occurs at 900 K [11].

On the other hand, some of the Ti-PGM compounds that have been investigated were previously found to have a highly stable B2 phase to room temperature, with no sign of martensitic transformation occurring, thus no shape memory behavior was observed. Amongst those are TiOs [12] and TiRu [13–16], which showed similar behavior to TiFe [17, 18]. So far, it is not known, at least from experimental studies, if these compounds may undergo martensitic transformation at cryogenic temperatures. However, such useful information can be generated using DFT computational tools. Moreover, from scientific point of view, it is very important to establish factors that suppress MT.

It is clear that the incorporation of PGMs in Ti influences both the crystal and electronic structure and consequently, the stability of austenite phase and formation of martensite phases. These phases are thoroughly studied as they are key in the design and functionality of SMAs for specific applications [19]. Most researchers have used computational calculations to gain some insight into the underlying shape memory properties of TiPt [20], TiPd [21] and TiRh [10]. Computational studies usually include structural, elastic and electronic properties. First principle calculations have been proven to be a useful and reliable tool for studying ground-state properties of these compounds [22–25].

In addition to structural, elastic and electronic properties, there has been an increase in the use of lattice vibration properties to predict the dynamic stability for a particular phase by calculating phonon dispersions. For example, Haskins et al. [26] recently used phonon dispersion calculations to resolve discrepancies associated with determination of ground-state phases for TiNi [26, 27]. Also, phonon dispersion calculations have been used by several researchers in order to determine the lattice vibration properties and predict the phase transformation path for various alloy systems [10, 20, 27].

Although DFT predictions may not agree perfectly with experimental observations, the accuracy of predicted results can still be scrutinized against the available experimental data in order to validate the accuracy of DFT calculations. In predicting phase stability, there is a link between the predicted thermodynamic, electronic and mechanical and lattice dynamic stability, which in turn can be used to predict the expected phases at a particular temperature. Some researchers [26, 28, 29] often report these stabilities separately and although that is acceptable, we observed that these stabilities can be used in connection to gain clear understanding of phase transformations for the investigated Ti-PGM compounds.

Furthermore, in order to improve shape memory properties or induce MT in TiOs and TiRu, factors influencing the shape memory behavior must be understood and this can be done by studying the predicted mechanical, electronic and dynamic stabilities. Thus, the main aim of this chapter is to demonstrate the versatile use of *ab-initio* methods based on DFT to track the shape memory behavior of these well-known Ti-PGM compounds by evaluating their ground-state stability of the high-temperature austenite phase (B2) with reference to mechanical, electronic and dynamical stability. This will assist the reader in identifying the underlying fundamental factors that drive the existence of shape memory behavior and further give an insight towards the design of new and improvement of existing SMAs.

2. Computational methodology

2.1 Density functional theory (DFT)

Computational modeling techniques offer an alternative way of investigating the properties of materials using computers, whereby the simulator builds a model of a real system and explores its behavior. One of the computational techniques that have received immense attention over the last few decades is the first-principles calculations, also known as *ab-initio* calculations. This interest is attributed to significant usefulness and insightfulness of its calculated data in the materials design.

First-principles methods are based on DFT formalism in which properties of materials, that is, the values of the fundamental constants and the atomic numbers of the atoms present can be calculated using the Schrödinger equation. Due to its improved accuracy achieved over the past years in predicting properties of real solids, an *ab-initio* approach is adopted in the current work to predict the ground-state and structural properties of several B2 intermetallic compounds at equiatomic compositions. Computing total energies of any system is a necessary starting point for first-principle calculations.

DFT is a quantum-mechanical method used for calculating ground-state properties of condensed matter systems without dealing directly with many electron states [30]. It was first formulated by Hohenberg and Kohn in 1964 [31] and then secondly developed by Kohn and Sham in 1965 [32]. DFT has helped in the development of independent-particle methods that take into account the particle's correlations and interactions. Hohenberg-Kohn demonstrated the first theorem that the ground-state properties of a many-electron system are uniquely determined by an electron density that depends on only three spatial coordinates [30].

$$E = E\left[n\left(\vec{r}\right)\right] \tag{1}$$

Where E is the total energy and n is the density. Within the Kohn-Sham scheme [30], consideration of an interacting electron gas moving in an external potential $v_e(r)$, as a variational principle leads to the effective single-electron Schrödinger equations,

$$\left\{-\nabla^2 + v\left([n]; \vec{r}\right)\right\}\psi_j\left(\vec{r}\right) = \in_j\psi_j\left(\vec{r}\right) \tag{2}$$

Kohn-Sham electrons in an effective potential, v_{eff} for a system of non-interacting, is solved as follows:

$$V_{eff}\left(\vec{r}\right) = v\left(\vec{r}\right) + \int \frac{n\left(\vec{r}'\right)}{|\vec{r} - \vec{r}'|} d\vec{r}' + \frac{\delta E_{xc}\left[n\left(\vec{r}\right)\right]}{\delta n\left(\vec{r}\right)} \tag{3}$$

Where $v\left(\vec{r}\right)$ is the external potential and $E_{xc}\left[n\left(\vec{r}\right)\right]$ is the exchange-correlation density functional [30].

2.2 Approximations to exchange-correlation functional

The two main types of exchange-correlation functionals used in DFT are the local density approximation (LDA) [33] and the generalized gradient approximation (GGA) [34], which have been discussed in the sub-sections below.

2.2.1 Local density approximation

The local density approximation (LDA) is an approximation in which the exchange-correlation (XC) energy functional in density functional theory (DFT) depends upon the value of the electronic density at each point in space. It was first discovered by Kohn and Sham, which is expressed in the Eq. (4):

$$E_{xc}^{LDA}[n(r)] = \int drn(r)\varepsilon_{xc}(n(r)) \tag{4}$$

Where $\varepsilon_{XC}(n)$ is the exchange-correlation energy per electron in a uniform electron gas of density n [33]. The uniform electron gas represents a group of systems of interacting electrons with an arbitrary spatially constant density n, which acts as a parameter. The local density approximation quantity is known for the limit of high density and can be calculated accurately at densities of interest by the use of Monte Carlo methods. LDA has been proven to give accurate results for many atomic, molecular and crystalline interacting electron systems, even though in these systems, the density of electrons is not slowly varied.

2.2.2 Generalized gradient approximation

The GGA is known to be a semi-local approximation, which means that there is no use of local density $n(r)$ value but its gradient $\nabla n(r)$. Perdew and Wang developed generalized gradient approximation (GGA), which is based on a real-space cut-off of the spurious long-range components for the second-order gradient expansion for the exchange-correlation hole [34]. GGA improves total energies, atomization energies, energy barriers and also the difference in structural energies. GGA takes the form:

$$E_{xc}^{GGA}\left[n\left(\vec{r}\right)\right] = \int \varepsilon_{xc}^{GGA}\left(n\left(\vec{r}\right), |\nabla n\left(\vec{r}\right)|\right)n\left(\vec{r}\right)d\vec{r} \tag{5}$$

There are several GGA-based functionals, that is, the PBE [35], PBEsol [36], RPBE [37], BLYP [38] and AM05 [39]. Other known GGA-based functionals are meta-GGA [40], hyper-GGA and generalized random phase approximation.

In this chapter, the GGA-PBE [35] functional was used to optimize the $Ti_{50}M_{50}$ (M = PGMs, Ni, Fe, Au) systems as it provides accurate parameters for these materials.

2.3 Computational code and implementation

2.3.1 CASTEP code

In this book chapter, the plane-wave Cambridge Serial Total Energy Package (CASTEP) [41, 42] code was used to investigate the properties of the B2 structures.

CASTEP is a module embedded within the Materials Studio software package. It is a first-principle quantum mechanical code based on DFT formalism, used for performing electronic structure calculations. It can be used to simulate a wide range of materials, including crystalline solids, surfaces, molecules, liquids and amorphous materials; the properties of any material that can be thought of as an assembly of nuclei and electrons can be calculated with the only limitation being the finite speed and memory of the computers being used. This approach to simulation is extremely ambitious given that the aim is to use no experimental data but to rely purely on quantum mechanics.

Aiming to calculate any physical property of the system from first principles, the basic quantity is the total energy from which many other quantities are derived. For example, the derivative of total energy concerning atomic positions results in the forces and the derivative concerning cell parameters gives stresses. To do this, the total-energy code, on CASTEP code, performs a variational solution to the Kohn-Sham equations by using a density mixing scheme to minimize the total energy and also conjugate gradients to relax the ions under the influence of the Hellmann-Feynman forces. CASTEP uses fast Fourier transforms (FFTs) to provide an efficient way of transforming various entities (wave functions and potentials) from real to reciprocal space and back, as well as to reduce the computational cost and memory requirement for operating with the Hamiltonian on the electronic wave functions, a plane-wave basis for the expansion of the wave functions. These are then used to perform full geometry optimizations [41, 42].

A summary of the methodology for electronic structure calculations as implemented in CASTEP is as follows: a set of one-electron Schrödinger (Kohn-Sham) equations are solved using the plane-wave pseudopotential approach. The wave functions are expanded in a plane wave basis set defined by the use of periodic boundary conditions and Bloch's Theorem. The electron-ion potential is described employing *ab initio* pseudopotentials within both norm-conserving and ultrasoft formulations [41, 42].

Direct energy minimisation schemes are used to obtain self-consistent electronic wave functions and corresponding charge density. In particular, the conjugate gradient and density mixing schemes are implemented. Also, the robust electron ensemble DFT approach can be used for systems with partial occupancies, in particular, metals [41, 42].

2.3.2 Implementation

Figure 1 illustrates the equiatomic B2 crystal geometry used to carry out all the calculations reported in this chapter.

The resulting geometry-optimized crystal structure was used to carry out all the calculations, including structural, thermodynamic and elastic properties, of all considered compounds. Only valence electrons were considered through the use of ultrasoft pseudopotentials [41, 42]. All of our calculations were performed with pseudo-potentials in generalized gradient approximation (GGA) [32] refined by Perdew, Burke and Ernzerhof (PBE) [35]. Before any calculation could be performed, a convergence test was also conducted in this code to determine the suitable cut-off energy and k-point mesh parameter for systems. A plane wave cut-off energy of 500 eV was found to be sufficient enough to converge the total energy of the systems. The Brillouin zone (BZ) sampling was performed using the k-point mesh of 13x13x13 according to the Monkhorst–Pack method [43]. A full geometry optimization was performed to determine the ground-state parameters for the binary systems.

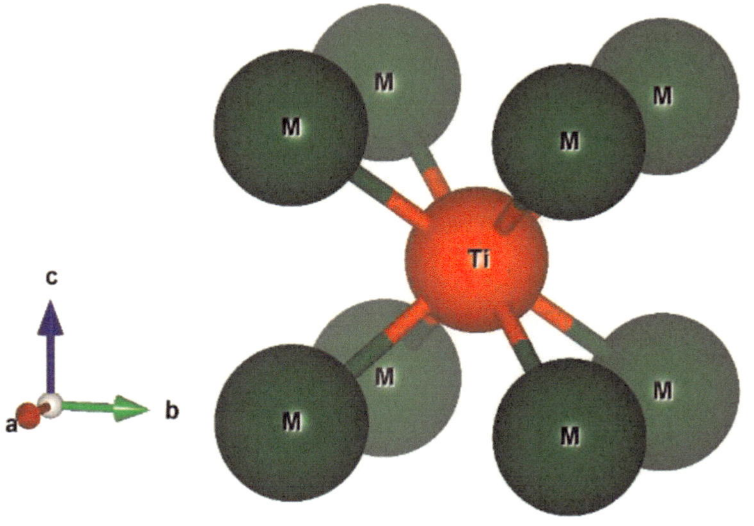

Figure 1.
Schematic representation of B2 crystal structure of $Ti_{50}M_{50}$ (M = Ru, Rh, Pd, Os, Ir, Pt, Ni, Fe, Au) intermetallic compounds.

To obtain the stable structure of $Ti_{50}M_{50}$ with minimum total energy, the Broyden-Fletcher-Goldfarb-Shanno (BFGS) minimization scheme [44] was performed in geometry optimization. The maximum ionic Hellmann-Feynman force was given to be 0.03 eV/Å. Furthermore, the elastic constants of a solid were calculated by an efficient strain–stress method through a linear least-square fit of first-principles calculation results. The maximum stress set is below 0.05 GPa. The phonons dispersion curves were also calculated.

2.4 Theoretical background on calculated properties

2.4.1 Thermodynamic stability

The heat of formation (ΔH_f) is the enthalpy change when one mole of a compound is formed from the constituent elements in their stable states and is essential in determining the structural stabilities of the different crystal structures. The heat of formation is estimated by the following expression:

$$\Delta H_f = E_C - \sum_i x_i E_i \tag{6}$$

where E_C is the calculated total energy of the compound and E_i is the calculated total energy of the constituent element in the compound. For a structure to be stable, the heat of formation must have the lowest negative value ($\Delta H_f < 0$). The heat of formation is used to determine the stability trend of the B2 systems.

2.4.2 Electronic stability

The electronic stability is determined by the density of states (DOSs), which refer to the occupancy and density of the electronic states in a crystalline solid. It is described by a function, g (E), as the number of electrons per unit volume and energy with electron energies near Fermi-level E_F. In the case of states with DOS being zero implies no state has been occupied (empty orbital). In general, a DOS is an average of all available spaces multiplied by the number of states occupied by the system. The local density of states (LDOSs) is a measure of variation due to distortion of the original system. LDOS can locally be non-zero if the DOS of an undisturbed system is zero due to the presence of local potential. In that case, the DOSs are the total number of states that are available in the system within the plane-wave framework of DFT. It is possible to calculate each orbital's contribution (partial DOS) to determine which orbitals are occupied or involved in bonding. The electronic behavior of a material is determined by the location of E_F within the DOS. Alloys' stability can be predicted using the DOS.

In the case of partial density of states (PDOSs), the states are attributed to the basic functions and then to the atoms constituting the unit cell. DOS is then calculated as the sum of atomic contributions. The DOS is calculated by using the following expression:

$$n(\varepsilon) = 2\sum_{n,k}\delta\left(\varepsilon - \varepsilon_n^k\right) = \frac{2}{V_{BZ}}\sum_n \int \delta\left(\varepsilon - \varepsilon_n^k\right)dk \tag{7}$$

where δ is the Dirac delta function and the k is integral extends over the BZ. The number of the electrons in the unit cell is given by:

$$\int_{-\infty}^{\varepsilon f} n(\varepsilon)d\varepsilon.$$

2.4.3 Mechanical stability

There are various criteria established to deduce the mechanical stability of crystals for different lattice crystals. Accuracy in determining the elasticity of a compound is vital in understanding its mechanical stability and elastic properties. The elastic constants depend on the type of lattice *i.e.* for the cubic, there are three (c_{11}, c_{12}, c_{44}) independent elastic constants [20, 45]. For example, applying two types of strains ($\varepsilon_1\varepsilon_4$) to the cubic system gives stresses relating to three elastic coefficients, this is a useful method for obtaining elastic constants. The mechanical stability condition for the cubic system as outlined in Ref. [45] is given as follows:

$$c_{44} > 0; \quad c_{11} > c_{12} \text{ and } c_{11} + 2c_{12} > 0 \tag{8}$$

According to Born-Huang's lattice dynamical theory [46, 47], the stability criterion for the elastic constants must be completely satisfied for the structure to be stable. The positive $C' = (1/2(c_{11}-c_{12}) > 0)$ indicates the mechanical stability of the crystal, otherwise, it is unstable.

2.4.4 Lattice dynamic stability

A phonon dispersion curve along a high symmetry direction is calculated by using interplanar force constants [45], as every plane perpendicular to this direction is displaced within an elongated supercell. Generally, lattice dynamics are analyzed with the *ab-initio* evaluation of forces on all atoms resulting from finite displacements of few atoms within otherwise perfect crystals. It is usually necessary to construct supercells of the appropriate size to ensure that interactions of the perturbation with all its translational symmetry equivalent copies are small. The techniques for selecting suitable supercells and atomic displacements, assembling force constant matrices from the calculated forces and calculating phonon dispersion relations *via* Fourier transform are well documented.

Using one of the 230 crystallographic space groups, phonon constructs a crystal structure, calculates the Hellmann-Feynman force constant, builds the dynamical matrix, diagonalizes it and calculates phonon dispersion relations [45].

In phonon dispersion calculations, polarization vectors and irreducible representations (Gamma points) of phonon modes are found, and the total and partial phonon densities are calculated. It plots the internal energy, free energy, entropy, heat capacity and tensor of mean square displacements (Debey-Waller factor). Phonons calculate the dynamical structure factor of coherent inelastic neutron scattering and incoherent doubly differential scattering in single crystal and polycrystalline systems [45].

The properties of phonons can be determined using a harmonic approximation with one fundamental quantity, the force constants matrix [45]:

$$D_{\mu\nu}(R - R') = \frac{\delta^2 E}{\partial u_\mu(R)\partial u_\nu}|_{u=0} \tag{9}$$

Where u represents the displacement of a given atom and E is the total energy in the harmonic approximation. This matrix of force constants can also be represented in reciprocal space and is known as a dynamic matrix:

$$D_{\mu\nu}(q) = \frac{1}{N_R}\sum_R D_{\mu\nu}(R)\exp(-iqR) \tag{10}$$

Classical equations of motion can be written as eigenvalue problems with each atomic displacement in the form of plane waves:

$$u(R,t) = \varepsilon e^{(qR-\omega(q)t)} \tag{11}$$

Where ε is the 3N-dimensional eigenvector of the eigenvalue problem is:

$$M\omega^2(q)\varepsilon = D(q)\varepsilon \tag{12}$$

The ω on the wave vector is well-known as the phonon dispersion. A guide to the basic theory of phonons has been described in detail by Born and Huang (1954) and Ashcroft and Mermin (1976) [48, 49]. In this chapter, we use the CASTEP code [41, 42] to calculate phonon dispersion curves and their density of states (PHDOS). It is well documented that compounds that radiate real (only positive) vibrational modes along high symmetry directions in the Brillouin zone are considered to be vibrationally stable with no possibility to undergo a martensitic phase transition at lower temperatures [10, 13]. On the other hand, compounds that radiate both positive and negative vibrational modes turn out to be vibrationally unstable with high chances to undergo a martensitic phase transition at lower temperatures, a primary feature of alloys with shape memory effect [20, 23, 50].

3. Results and discussion

3.1 Structural and thermodynamic stability

The stability of the investigated B2 compounds is first discussed on the basis of heat of formation. **Table 1** presents the ground-state lattice parameters and the heat of formation of the investigated B2 phases. The reported lattice parameters were obtained from the geometrically relaxed structures, whereby their respective volumes and unit cells were allowed to change to obtain their ground state.

The lattice parameters were found to be in good agreement with those reported previously by other authors [10, 12, 13, 19, 22, 54, 55]. The lattice parameters of the investigated $Ti_{50}M_{50}$ compounds increase with the position of the M atom along the groups or the periods of the periodic table of elements. This is in line with their respective densities and volume changes observed, as presented in **Table 1**.

Furthermore, **Table 1** also presents the thermodynamic stability of the investigated B2 compounds that were determined by calculating their respective heat of formation using Eq. (6). Heat of formation provide primary insight into the existence of the phase. All the investigated B2 phases reported in this research work were found to be negative ($\Delta H_F < 0$), an indication that they are all thermodynamically stable. Amongst the investigated B2 compounds, the heat of formation value of $Ti_{50}Ir_{50}$ was found to be the most thermodynamically stable indicated by the lowest value of −0.913 eV/atom. The reported heat of formation results were found to be in accordance with other data in literature [51, 52, 55–58].

3.2 Electronic stability

Figure 2 shows the total density of states (tDOS) of all the investigated B2 compounds reported in this chapter. It is noted that the DOS values $g(E)$ of the investigated compounds were found to be non-zero across the Fermi level (E_F) indicating that all the investigated B2 compounds were mainly characterized by metallic bonds.

Crystal structures	Lattice parameters, a (Å)		Volume ($Å^3$)	Density ($g.cm^3$)	Heat of formation, $-\Delta H_F$ (eV/atom)	
	This work	Literature			This work	Literature
$Ti_{50}Ru_{50}$	3.08	3.07, 3.09	29.21	8.47	0.750	0.798, 0.743, 0.770 [14, 51]
$Ti_{50}Rh_{50}$	3.12	3.13	30.41	8.23	0.852	0.741, 0.749, 0.715 [51]
$Ti_{50}Pd_{50}$	3.17	3.18	31.74	8.07	0.508	0.519, 0.530 [14, 21, 51]
$Ti_{50}Os_{50}$	3.09	3.09, 3.10	29.58	13.37	0.705	0.710, 0.714, 0.683 [51]
$Ti_{50}Ir_{50}$	3.12	3.12	30.35	13.14	0.913	0.876, 0.847, 0.845 [51]
$Ti_{50}Pt_{50}$	3.18	3.19, 3.21	32.18	12.54	0.885	0.824, 0.795 [51]
$Ti_{50}Fe_{50}$	2.95	2.99	25.73	6.70	0.745	0.321 [52]
$Ti_{50}Ni_{50}$	3.01	3.02, 2.97	27.34	6.47	0.382	0.358 [53]
$Ti_{50}Au_{50}$	3.27	3.25	34.96	11.63	0.368	0.442 [51]

Table 1.
Lattice parameters and heat of formation of the investigated binary B2 compounds.

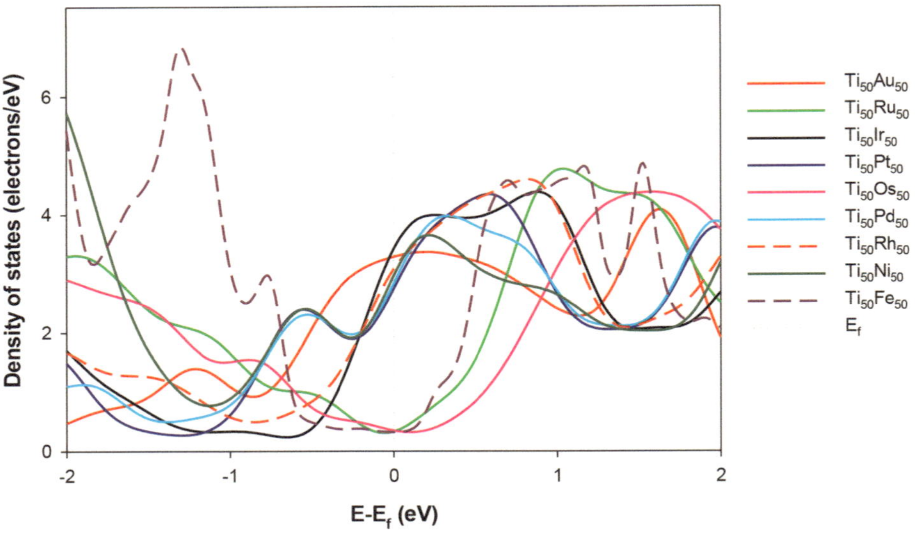

Figure 2.
The total density of states (tDOS) of the nine investigated austenite compounds.

Generally, the density of states' spectra of CsCl-type B2 compounds consists of two peaks [25] that are separated by a pseudogap (deep valley) at the Fermi level (E_F). With the lower energy peak representing the anti-bonding region and the higher energy peak representing the bonding region. Therefore, the position of E_F at the pseudogap provides insight into the phase stability of a compound at the ground state. Such that, if E_F is found to fall at the centre of the pseudogap, that signifies phase stability and if E_F falls at the peak or shoulder of the bonding region. **Figure 2** shows that $Ti_{50}Fe_{50}$, $Ti_{50}Ru_{50}$ and $Ti_{50}Os_{50}$ will retain their stable high-temperature austenite phase as their ground-state as their E_F cuts at the centre of the deep valley, while the other investigated B2 compounds ($Ti_{50}M_{50}$, M = Ni, Rh, Pd, Ir, Au and Pt) will certainly undergo a phase transition from the high-temperature B2 to unstable phases at 0 K because their pseudogap shifted towards the anti-bonding region and enabled phase transition.

The aforesaid provides insight information about the phase transition of the investigated compounds from high-temperature austenite (B2) to low-temperature martensite phase, a gauge for shape memory effect observed on alloys with shape memory properties.

3.3 Dynamic phase stability

The phonon-dispersion curves are used to gain insight into the underlying lattice vibrations that may influence the ability of the crystal to transform to martensite phase on cooling. The information on phonons is very useful for accounting variety of properties and behaviors of crystalline materials, such as thermal properties, phase transition, and superconductivity [59].

As detailed in Section 2.4.4, B2 compounds that show only positive frequencies remain stable with no prospect of undergoing martensitic phase transition, while those that show both positive and negative frequencies are prone to become unstable and undergo martensitic phase transition at lower temperatures.

Figure 3 represents the sets of phonon-dispersion curves of the investigated compounds. It can be shown that $Ti_{50}Fe_{50}$, $Ti_{50}Ru_{50}$ and $Ti_{50}Os_{50}$ consist of only the real positive vibrational modes, while the rest of the investigated compounds were found to consist of both positive and negative vibrational modes. This observed behavior agrees well with the DOS results reported in the previous section as well as the results reported by other authors [20, 23, 50].

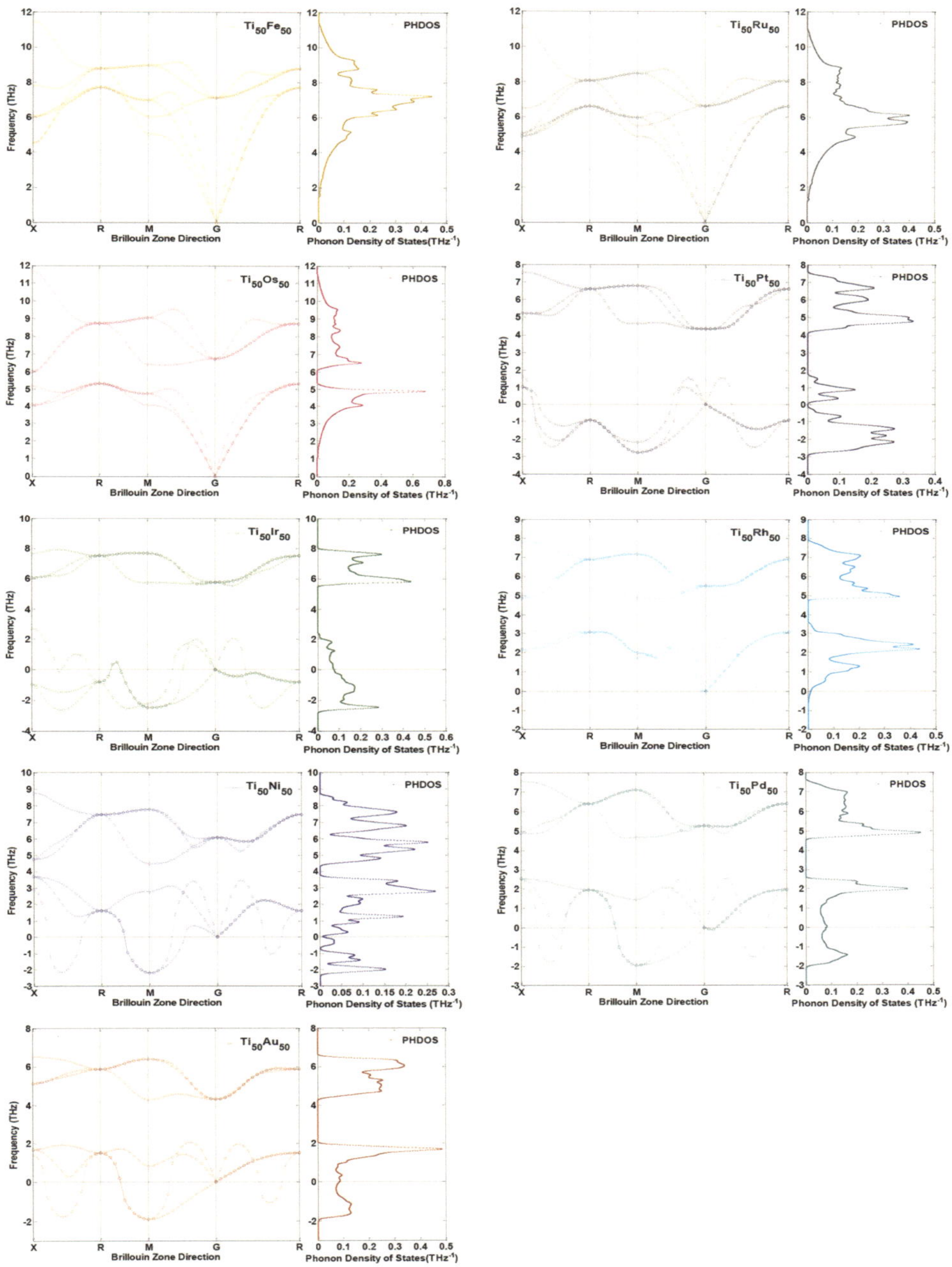

Figure 3.
Phonon-dispersion curves of the investigated B2 phases plotted along selected Brillouin zone directions.

Furthermore, **Figure 3** gives the calculated B2 phonons density of states (PHDOS). It is evident that the vibrational stable B2 compounds ($Ti_{50}M_{50}$, M = Fe, Ru and Os) consist of a valence gap between the positive and negative frequency modes. Such observed behavior renders the aforementioned compounds to be brittle-metal-compounds, a kind of semi-conductive material properties. While those austenite phases that were found to be vibrationally unstable ($Ti_{50}M_{50}$, M = Ni, Rh, Pd, Ir, Pt and Au) show a clear interaction of valence and conduction electrons with a non-zero DOS.

3.4 Mechanical stability

Elastic constants are the primary output parameters of most DFT codes, and they determine the response of the crystal to external forces. They are crucial in predicting the mechanical properties of the crystal structure and their subsequent modulus of elasticity.

Table 2 presents the calculated elastic constants of the investigated alloys reported in this chapter. Furthermore, the trend of the elastic constants plotted against different B2 compounds is shown in **Figure 4**.

Based on the mechanical stability criterion for cubic crystals as outlined in Section 2.4.3, as well as very high value of C′, $Ti_{50}Fe_{50}$, $Ti_{50}Ru_{50}$ and $Ti_{50}Os_{50}$ completely

Parameters	TiFe	TiNi	TiRu	TiRh	TiPd	TiOs	TiIr	TiPt	TiAu
C_{11}	357.43	206.68	396.10	166.15	148.46	466.03	−55.28	141.65	131.78
C_{12}	101.88	138.67	122.55	206.86	163.98	140.21	376.01	203.15	139.44
C_{44}	71.64	47.08	82.72	59.41	51.51	125.20	79.08	48.27	43.08
C'	127.78	34.00	136.78	−19.85	−7.76	162.91	−215.65	−30.75	−3.83

Table 2.
The calculated elastic constants (C_{ij}) of the investigated $Ti_{50}M_{50}$ compounds.

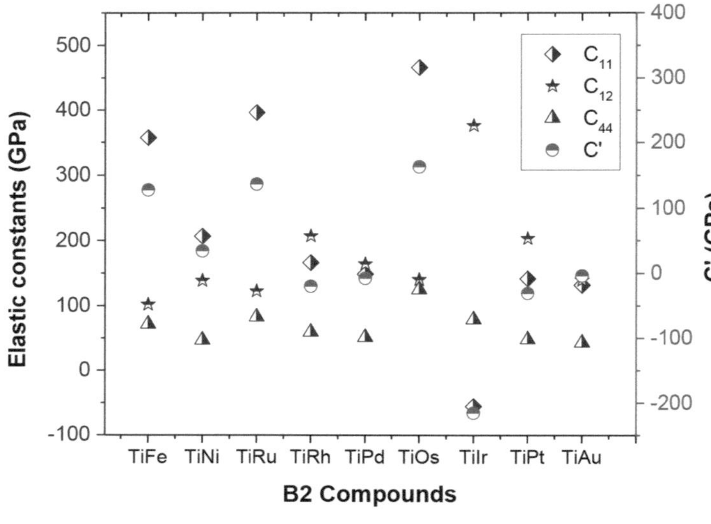

Figure 4.
The calculated elastic constants (GPa) of the investigated compounds.

satisfy the criterion, rendering the austenite phase to be mechanically stable in these compounds, thus unlikely to undergo any phase transition at lower temperatures.

Although a similar stability criterion was observed ($C' > 0$) for $Ti_{50}Ni_{50}$, which is a well-known SMA [14, 60], it should be noted that its C_{44} was found to be larger than C', resulting in anisotropic factor A greater than 1. This is another important elastic property factor worth attention in design of SMAs and can be assessed on the relationship between the tetragonal shear modulus (C') and its corresponding monoclinic shear constant (C_{44}). It is reported that the mechanical instability of a high-temperature phase decreases when nearing the martensitic transition temperature during heating [61], this is the case where C' becomes close to zero. Furthermore, if $C' < 0$, the corresponding B2 phase is unstable at 0 K, signaling a great potential for this high-temperature phase to undergo a martensitic phase transition on cooling, yielding SME character. Since this MT occurs at much higher temperatures, such B2 alloys are of interest for high-temperature structural applications. **Table 2** shows that $Ti_{50}Os_{50}$ and $Ti_{50}Ir_{50}$ are the most and least mechanically stable amongst the investigated B2 compounds, and the same results are graphically presented as shown in **Figure 4**.

4. Conclusions

The thermodynamic, electronic, lattice dynamic and mechanical stability of the investigated B2 compounds have been calculated using a first-principles approach and are reported in this chapter. The calculated heat of formation, elastic constants, densities of states and vibrational properties have provided a much-needed clarity in determining the differences in stabilities of various Ti-PGM B2 compounds. This was made possible by choosing the correct model, resulting in reliable data that compares well with results reported in literature.

From all the calculated properties, it was shown that the higher-temperature austenite phase of $Ti_{50}Fe_{50}$, $Ti_{50}Ru_{50}$ and $Ti_{50}Os_{50}$ remains stable with no prospect of martensitic phase transition that is associated with materials with shape memory effect.

This was evident by the electronic and dynamic stability of the investigated compounds. Furthermore, their corresponding tDOS spectra were found to coincide with the pseudogap at the E_F, rendering electronic stability with no phase transition. This was further substantiated by phonon curves, which possessed only positive vibrational frequency indicating that they are not likely to undergo a phase transition. The predicted mechanical stability of the $Ti_{50}Fe_{50}$, $Ti_{50}Ru_{50}$ and $Ti_{50}Os_{50}$ B2 compounds was found to be very high, strongly signaling absence of any possible phase transformation.

On the other hand, using the electronic and elastic properties, this work has shown that other B2 compounds ($Ti_{50}M_{50}$, M = Ni, Rh, Pd, Ir, Pt and Au) considered in this study are unstable at 0 K, thus predicting the possibility to undergo phase transition to martensite phase on cooling, resulting in shape memory effect character if the transformation is diffusionless.

Acknowledgements

This book chapter is published with the permission of MINTEK. The authors would like to thank the Advanced Metals Initiative (AMI) of the Department of Science and Innovation (DSI) for financial support. The gratitude is also extended to the Centre for High-Performance Computing (CHPC) in Cape Town for allowing us to carry out the calculations using their remote computing resources.

Author details

Ramogohlo Diale[1], Duduzile Nkomo[1,2]*, Bongani Ngobe[1,3] and Maje Phasha[1]

1 Advanced Materials Division, Mintek, Johannesburg, South Africa

2 Department of Materials Science and Metallurgical Engineering, University of Pretoria, Pretoria, Hatfield, South Africa

3 School of Physics, University of the Witwatersrand, Johannesburg, South Africa

*Address all correspondence to: dudunk@mintek.co.za

References

[1] Jani JM, Leary M, Subi A, Gibson MA. A review of shape memory alloy research, applications and opportunities. Materials and Design. 2014;**56**:1078-1113

[2] Biggs T, Cortie MB, Witcomb MJ, Cornish LA. Martensitic transformations, microstructure, and mechanical workability of TiPt. Metallurgical and Materials Transactions A. 2001;**32**:1881-1886

[3] Yamabe-Mitarai Y, Hara T, Miura S, Hosoda H. Shape memory effect and pseudoelasticity of TiPt. Intermetallics. 2010;**18**:2275-2280

[4] Otsuka K, Oda K, Ueno Y, Piao M, Ueki T, Horikawa H. The shape memory effect in a Ti50Pd50 Alloy. Scripta Metallurgica et Materialia. 1993;**29**: 1355-1358

[5] Solomon VC, Nishida M. Martensitic transformation in Ti-Rich Ti–Pd shape memory alloys. Materials Transactions. 2002;**43**:908-915

[6] Hisada S, Matsuda M, Yamabe-Mitarai Y. Shape change and crystal orientation of B19 martensite in equiatomic TiPd alloy by isobaric test. Meta. 2020;**10**:3754

[7] Yamabe-Mitarai Y, Hara T, Phasha MJ, Ngoepe PE, Chikwanda HK. Phase transformation and crystal structure of IrTi. Intermetallics. 2012;**31**: 26-33

[8] Yi SS, Chen BH, Franzen HF. Phase transitions in RhTi. Journal of the Less-Common Metals. 1988;**143**:243

[9] Szajek A. The electronic and superconducting properties of ordered Ti-Rh alloys. Journal of Physics. Condensed Matter. 1991;**3**:1089

[10] Si L, Jiang Z-Y, Zhou B, Chen W-Z. First-principles study of martensitic phase transformation of TiRh alloy. Physica B. 2012;**407**:347-351

[11] Donkersloot HC, Van Vucht JHN. Martensitic transformations in Gold–Titanium, Palladium–Titanium, and Platinum–Titanium alloys near the equiatomic composition. Journal of the Less-Common Metals. 1970;**20**: 83-91

[12] Murray JL. Ti Os-Ti (Osmium-Titanium) system. Bulletin of Alloy Phase Diagrams. 1982;**3**:212-215

[13] Kong Z, Duan Y, Peng M, Qu D, Bao L. Theoretical predictions of thermodynamic and electronic properties of TiM (M=Fe, Ru and Os). Physica B: Condensed Matter. 2019;**573**: 13-21

[14] Xing W, Chen XQ, Li D, Li Y, Fu CL, Meschel SV, et al. First-principles studies of structural stabilities and enthalpies of formation of refractory intermetallics: TM and TM3 (T=Ti, Zr, Hf; M=Ru, Rh, Pd, Os, Ir, Pt). Intermetallics. 2012;**28**:16-24

[15] Murray JL. The Ru-Ti (Ruthenium-Titanium) system. Bulletin of Alloy Phase Diagrams. 1982;**3**:216-221

[16] Jain E, Pagare G, Chouhan SS, Sanyal PS. Electronic structure, phase stability and elastic properties of ruthenium based four intermetallic compounds: Ab-initio study. Intermetallics. 2014;**59**:79-84

[17] Cacciamani G, De Keyzer J, Ferro R, Klotz UE, Lacaze J, Wollants P. Critical evaluation of the Fe–Ni, Fe–Ti and Fe–Ni–Ti alloy systems. Intermetallics. 2006;**14**:1312-1325

[18] Murray JL. The Fe-Ti (Iron-Titanium) system. Bulletin of Alloy Phase Diagrams. 1981;**2**:320-334

[19] Wadood A, Yamabe-Mitarai Y. Recent research and developments related to near-equiatomic titanium-platinum alloys for high-temperature applications. Platinum Metals Review. 2014;**58**:61-67

[20] Mahlangu R, Phasha MJ, Chauke HR, Ngoepe PE. Structural, elastic and electronic properties of equiatomic PtTi as potential high-temperature shape memory alloy. Intermetallics. 2013;**33**:27-32

[21] Chen XQ, Fu CL, Morris JR. The electronic, elastic, and structural properties of Ti–Pd intermetallics and associated hydrides from first principles calculations. Intermetallics. 2010;**18**: 998-1006

[22] Bao W, Liu D, Duan Y, Peng M. First-principles predictions of anisotropies in elasticity and sound velocities of CsCl-type refractory intermetallics: TiTM, ZrTM and HfTM (TM = Fe, Ru, Os). Philosophical Magazine. 2019;**99**:2681-2702

[23] Baloyi ME, Modiba R, Ngoepe PE, Chauke HR. Structural, elastic and electronic properties of titanium-based shape memory alloys. In: The Proceedings of South African Institute of Physics 2018. South Africa: SAIP; 2018. pp. 8-13. ISBN: 978-0-620-85406-1

[24] Hart GLW, Curtarolo S, Massalski TB, Levy O. Comprehensive search for new phases and compounds in binary alloy systems based on platinum-group metals, using a computational first-principles approach. Physical Review X. 2013;**3**: 041035

[25] Zhang JM, Guo GY. Electronic structure and phase stability of three series of B2 Ti-transition-metal compounds. Journal of Physics: Condensed Matter. 1995;**7**:6001-6017

[26] Haskins JB, Lawson JW. Ab initio simulations of phase stability and martensitic transitions in NiTi. Physical Review B. 2016;**94**(21):1-10

[27] Haskins JB, Hessam M, Honrao SJ, Sandoval LA, Lawson JW. Low-temperature mechanical instabilities govern high-temperature thermodynamics in the austenite phase of shape memory alloy constituents: Ab Initio simulations of NiTi, NiZr, NiHf, PdTi, and PtTi. Acta Materialia. 2021; **212**:1-13

[28] Vishnu KG, Alejandro S. Phase stability and transformations in NiTi from density functional theory calculations. Acta Materialia. 2010;**58**(3): 745-752

[29] Abdoshahi N, Dehghani M, Hatzenbich L, Spoerk-Erdely P, Ruban AV, Musi M, et al. Structural stability and mechanical properties of TiAl+Mo alloys: A comprehensive ab initio study. Acta Materialia. 2021;**221**: 1-14

[30] Mattson AE, Schultz PA, Desjarlais MP, Mattsson TR, Leung K. Designing meaningful density functional theory calculations in materials science-a primer. Modelling and Simulation in Materials Science and Engineering. 2005; **13**:1-32

[31] Hohenberg P, Kohn W. Introduction to density functional theory. Physical Review B. 1964;**136**:864

[32] Kohn W, Sham LJ. Self-consistent equations including exchange and correlation effects. Physical Review A. 1965;**140**:1133

[33] Hedin L, Lundqvist BI. Explicit local exchange-correlation potentials. Journal of Physics C. 1971;**4**:2064-2082

[34] Perdew JP, Wang Y. Accurate and simple analytic representation of the electron-gas correlation energy. Physical Review B. 1992;**45**:13244-13249

[35] Perdew JP, Burke K, Ernzerhof M. Generalized gradient approximation made simple. Physical Review Letters. 1996;**77**:3865-3868

[36] Perdew JP, Ruzsinszky A, Csonka GI, Vy-drov OA, Scuseria GE, Constantin LA, et al. Restoring the density-gradient expansion forexchange in solids and surfaces. Physical Review Letters. 2008;**100**:136406

[37] Hammer B, Hansen LB, Nørskov JK. Improved adsorption energetics within density-functionaltheory using revised Perdew-Burke-Ernzerhof functionals. Physical Review B. 1999;**59**:7413

[38] Gill PMW, Johnson BG, Pople JA, Frisch MJ. The performance of the Becke-Lee-Yang-Parr (B-LYP) density functional theory with various ba-sis sets. Chemical Physics Letters. 1992;**197**: 499

[39] Armiento R, Mattsson AE. Functional designed toinclude surface effects in self-consistent density functional theory. Physical Review B. 2005;**72**:085108

[40] Tao J, Perdew JP, Staroverov VN, Scuseria GE. Climbing the density functional ladder: nonempirical meta-generalized gradient approximation designed for molecules and solids. Physical Review Letters. 2003;**91**:146401

[41] Segall MD, Philip JDL, Probert MJ, Pickard CJ, Hasnip PJ, Clark SJ, et al. First-principles simulation: Ideas, illustrations and the CASTEP code. Journal of Physics D: Condensed Matter. 2002;**14**:2717-2744

[42] Clark SJ, Segall MD, Pickard CJ, Hasnip PJ, Probert MI, Refson K, et al. First principles methods using CASTEP. Zeitschrift für Kristallographie. 2005; **220**:567-570

[43] Monkhorst HJ, Pack JD. Special points for Brillouin-zone integrations. Physical Review B. 1996;**13**:5188-1592

[44] Fischer T, Almlof J. General methods for geometry and wave function optimization. Journal of Physical Chemistry. 1992;**96**:9768-9774

[45] Mehl MJ, Klein BM. First-principles calculation of elastic properties. In: Fleischer JW, editor. Intermetallic Compounds – Principles and Practice. Vol. 1. London: Principles, John Wiley and Sons; 1994. pp. 195-210

[46] Gomisa O, Manjónb FJ, Rodríguez-Hernándezc P, Muñozc A. Elastic and thermodynamic properties of α-Bi2O3 at high pressures: Study of mechanical and dynamical stability. Journal of Physics and Chemistry of Solids. 2019;**124**: 111-120

[47] Wagner MF-X, Windl W. Lattice stability, elastic constants and macroscopic moduliof NiTi martensites from first principles. Acta Materialia. 2008;**56**:6232-6245

[48] Born M, Huang K. Dynamical Theory of Crystal Lattices. Oxford: Oxford University Press; 1954

[49] Ashcroft NW, Mermin ND. Solid State Physics. Philadelphia: Saunders College; 1976

[50] Diale RG, Modiba R, Ngoepe PE, Chauke HR. The effect of Ru on

Ti50Pd50 high temperature shape memory alloy: A first-principles study. MSR Advances. 2019;**4**:2419-2429

[51] Semenova EL. Alloys with the shape memory effect in systems of d-transition metals containing metals of the platinum group. Powder Metallurgy and Metal Ceramics. 1997;**36**:394-404

[52] Gachon JC, Selhaoui N, Aba B, Hertz J. Comparison between measured and predicted enthalpies of formation. Journal of Phase Equilibria. 1992;**13**: 506-511

[53] Mizuno M, Akari H, Shirai Y. Compositional dependence of structures of NiTi martensite from first principles. Acta Materialia. 2015;**95**: 184-191

[54] Philip TV, Beck PA. CsCI-type ordered structures in binary alloys. Journal of Metals. 1957;**1957**: 1269-1271

[55] Jung J, Ghoshand G, Olson GB. A comparative study of precipitation behavior of Heusler phase (Ni_2TiAl) from B2-TiNi in Ni–Ti–Al and Ni–Ti–Al–X (X = Hf, Pd, Pt, Zr) alloys. Acta Materialia. 2003;**51**:6341-6357

[56] John R, Ruben H. Theoretical investigations of Ti-based binary shape memory alloys. Materials Sciences and Applications. 2011;**2**:1355-1366

[57] Topor L, Kleppa OJ. Thermochemistry of the intermetallic compounds RuTi, RuZr, and RuHf. Metallurgical Transactions A. 1988;**9A**: 1061-1066

[58] Guo Q, Kleppa OJ. Standard enthalpies of formation of some alloys formed between group IV elements and group VIII elements, determined by high-temperature direct synthesis calorimetry II. Alloys of (Ti, Zr, Hf) with (Co, Ni). Journal of Alloys and Compounds. 1998;**269**:181-186

[59] Togo A, Tanaka I. First principles phonon calculations in materials science. Scripta Materialia. 2015;**108**:1-5

[60] Buehler WJ, Gilfrich JW, Wiley RC. Effect of low temperature phase changes on the mechanical properties of alloys near composition TiNi. Journal of Applied Physics. 1963;**34**:1475

[61] Sedlák P, Janovská M, Bodnárová L, Heczko O, Seiner H. Softening of shear elastic coefficients in shape memory alloys near the martensitic transition: A study by laser-based resonant ultrasound spectroscopy. Meta. 2020;**10**:1-13

Chapter 2

DFT and TDDFT Calculations of Ground and Excited States of Photoelectron Emission

Brahim Ait Hammou, Abdelhamid El Kaaouachi, El Hassan Mounir, Hamza Mabchour, Abdellatif El Oujdi, Adil Echchelh, Said Dlimi and Driss Ennajih

Abstract

The Density-Functional Theory (DFT) is a reformulation of the quantum study of a correlated N-body system into a simpler system with independent equations being solved iteratively. The DFT considers only ground states of the systems. The extension to the time-dependent case of this theory is the Time-Dependent Density-Functional Theory (TDDFT) that also takes into account the excited states of the system. These calculations are very interesting in photonics areas. In fact, the interaction between electrons and light in the vicinity of solid surfaces and nanostructures is important as pathway to integrate photonics and electronics. The capability to couple light and electrons in purposefully designed device depends on the capability of creating such devices and the understanding of the underlying science.

Keywords: DFT, TDDFT, solid surfaces, ground states, excited states

1. Introduction

In the fields of photonics, researchers use the dielectric function and the electronic density in their calculations and investigations [1–8] to determine the physical and optical characteristics of materials such as noble metals (Au and Ag). In fact, the determination of the electronic structure of a material can give us all the physical and optical information about it. DFT and TDDFT are used to determine the electronic structure of materials such as noble metals in their ground and excited states. The purpose of these calculations is to determine the electron density $\rho\left(\vec{r}\right)$. Indeed, it informs us about all the physical properties of the studied system. We simplify this very complex calculation by using several approximations and theorems. We propose at the end a flowchart to carry out a computer simulation allowing to highlight the values of the electron density $\rho\left(\vec{r}\right)$.

2. DFT calculation steps

The DFT is based on wave Schrodinger equation

$$\hat{H}_T \psi\left(\{\vec{R}_I\},\{\vec{r_i}\},t\right) = i\frac{\partial}{\partial t}\psi\left(\{\vec{R}_I\},\{\vec{r_i}\},t\right) \tag{1}$$

We consider that the system contains N cores and M electrons.

$\hat{H}_T$: is the total Hamiltonian representing N cores and M electrons.

$\psi\left(\{\vec{R}_I\},\{\vec{r_i}\},t\right)$: is the wave function representing N cores and M electrons.

$\{\vec{R}_I\}$and $\{\vec{r_i}\}$: represent respectively the set of nuclear and electronic coordinates.

In the case of stationary processes, the Schrodinger becomes:

$$\hat{H}_T \psi\left(\{\vec{R}_I\},\{\vec{r_i}\}\right) = iE\psi\left(\{\vec{R}_I\},\{\vec{r_i}\}\right) \tag{2}$$

where E represent the energy of the system described by $\psi\left(\{\vec{R}_I\},\{\vec{r_i}\}\right)$

$$\hat{H}_T = \hat{T}_T + \hat{V}_T \tag{3}$$

where $\hat{T}_T$ is the total kinetic energy of the system operator, and $\hat{V}_T$ is the operator describing all Colombian interaction.

$\hat{H}_T$ can be written as:

$$\hat{H}_T = \hat{T}_n + \hat{T}_e + \hat{V}_{n-e} + \hat{V}_{e-e} + \hat{V}_{n-n} \tag{4}$$

where

$\hat{T}_n = \frac{-\hbar^2}{2}\sum_I \frac{\nabla^2 \vec{R}_I}{M_n}$: is the kinetic energy of N cores with mass M_n.

$\hat{T}_n = \frac{-\hbar^2}{2}\sum_i \frac{\nabla^2 \vec{r}_i}{m_e}$: is the kinetic energy of M electrons with mass m_e.

$\hat{V}_{n-e} = \frac{1}{4\pi\varepsilon_0}\sum_{i,j}\frac{e^2 Z_i}{\left|\vec{R}_i - \vec{r}_j\right|}$: is the core-electron attractive Colombian interaction.

$\hat{V}_{e-e} = \frac{1}{8\pi\varepsilon_0}\sum_{i\neq j}\frac{e^2}{\left|\vec{r}_i - \vec{r}_j\right|}$: is the electron-electron repulsive Colombian interaction.

$\hat{V}_{n-n} = \frac{1}{8\pi\varepsilon_0}\sum_{i\neq j}\frac{e^2 Z_i Z_j}{\left|\vec{R}_i - \vec{R}_j\right|}$: is the core-core repulsive Colombian interaction.

Eq. (4) takes into account N cores and M electrons. Some simplifications must be made:

- First approximation level: Born-Oppenheimer approximation. (see paragraph 2-1)
- Second approximation level: Hartree-Fock approximation. (see paragraph 2-2)
- Third approximation level: Inherent approximation to the resolution of equation. (see paragraph 2-3)

2.1 Adiabatic born-Oppenheimer approximation

It offers the opportunity to treat separately the cores and the electrons.
So we have

$$\Psi_R\left(\vec{R}, \vec{r}\right) = \Phi_R\left(\vec{R}\right)\Psi_R\left(\vec{r}\right) \tag{5}$$

where
$\Phi_R\left(\vec{R}\right)$: is the wave function describing the cores.
$\Psi_R\left(\vec{r}\right)$: is the wave function describing the electrons.
In this approximation, the interaction electron-phonon is neglected.
Using Eqs. (3) and (4), the new Hamiltonian becomes

$$\hat{H}_T = \hat{T}_e + \hat{V}_{n-e} + \hat{V}_{e-e} + \left(\hat{V}_{n-n} = \hat{V}_{ext} = Cst\right) \tag{6}$$

so

$$\hat{H}_T = \frac{-\hbar^2}{2}\sum_i \frac{\nabla^2 \vec{r}_i}{m_e} + \frac{1}{4\pi\varepsilon_0}\sum_{i,j}\frac{e^2 Z_i}{\left|\vec{R}_i - \vec{r}_j\right|} + \frac{1}{8\pi\varepsilon_0}\sum_{i\neq j}\frac{e^2}{\left|\vec{r}_i - \vec{r}_j\right|} + \left(\frac{1}{8\pi\varepsilon_0}\sum_{i\neq j}\frac{e^2 Z_i Z_j}{\left|\vec{R}_i - \vec{R}_j\right|} = Cst\right) \tag{7}$$

The nuclear kinetic energy term independent from electrons is canceled (equal to zero). The attractive potential energy (electron-core) term is removed and the repulsive potential energy (core-core) becomes a constant simply evaluated for determined geometry. Using the Hohenberg and Kohn Theorems (HK theorems) [9], the formulation of the Schrödinger equation can be now based on the electron density $\rho\left(\vec{r}\right)$. This is due to the two HK theorems. In fact, the first KH theorem indicated that the total energy of the system in the ground state is a single universal function of the electronic density.

$$E = E\left[\rho\left(\vec{r}\right)\right] \tag{8}$$

HK gives

$$E\left[\rho\left(\vec{r}\right)\right] = F_{HK}\left[\rho\left(\vec{r}\right)\right] + \int \hat{V}_{ext}\left(\vec{r}\right)\rho\left(\vec{r}\right)d\vec{r}, \tag{9}$$

where $F_{HK}\left[\rho\left(\vec{r}\right)\right]$ represent the HK universal functional and $\vec{V}_{ext}\left(\vec{r}\right)$ represent the external potential. The second HK theorem says that it is an analogue variational principle to the originally proposed in the approach of Hartree-Fock for the functional of the wave function ($\frac{\partial E[\psi]}{\partial \psi} = 0$), but this time applied to the electronic density functional.

$$\left.\frac{\partial E\left[\rho\left(\vec{r}\right)\right]}{\partial\rho\left(\vec{r}\right)}\right|_{\rho_0\left(\vec{r}\right)} = 0 \tag{10}$$

where $\rho_0\left(\vec{r}\right)$ is the exact electronic density of the system in the ground state. This theorem can be reformulated like: For a potential $\hat{V}_{ext}\left(\vec{r}\right)$ and with a number M of electrons, the total energy of the system reaches its minimum value when $\rho\left(\vec{r}\right)$ is corresponding to the exact density $\rho_0\left(\vec{r}\right)$ of the ground state.

Considering the two HK theorems, the resolution of Schrödinger equation consists in seeking the minimization of $E\left[\rho\left(\vec{r}\right)\right]$: $\frac{\partial E\left[\rho\left(\vec{r}\right)\right]}{\partial\rho\left(\vec{r}\right)} = 0$ by applying the constraint of the conservation of the total number of particles: $\int\rho\left(\vec{r}\right)d\vec{r} = M$.

This problem can be resolved using Lagrange multipliers

$$G\left[\rho\left(\vec{r}\right)\right] = \int\rho\left(\vec{r}\right)d\vec{r} - M \tag{11}$$

The conservation of the total number of particles constraint becomes $G\left[\rho\left(\vec{r}\right)\right] = 0$, and by introducing auxiliary function $A\left[\rho\left(\vec{r}\right)\right]$ with:

$$A\left[\rho\left(\vec{r}\right)\right] = E\left[\rho\left(\vec{r}\right)\right] - \mu G\left[\rho\left(\vec{r}\right)\right] \tag{12}$$

where μ is the Lagrange multiplier, the problem to resolve becomes

$$\partial A\left[\rho\left(\vec{r}\right)\right] = \int\frac{\partial A\left[\rho\left(\vec{r}\right)\right]}{\partial\rho\left(\vec{r}\right)}\partial\rho dr = 0 \tag{13}$$

$$\partial\left\{E\left[\rho\left(\vec{r}\right)\right] - \mu\left[\int\rho\left(\vec{r}\right)d\vec{r} - M\right]\right\} = 0$$

We must calculate the functional derivative of $A\left[\rho\left(\vec{r}\right)\right]$:

$$\frac{\partial A\left[\rho\left(\vec{r}\right)\right]}{\partial\rho\left(\vec{r}\right)} = \frac{\partial}{\partial\rho\left(\vec{r}\right)}\left\{E\left[\rho\left(\vec{r}\right)\right] - \mu\left[\int\rho\left(\vec{r}\right)d\vec{r} - M\right]\right\} = \frac{\partial E\left[\rho\left(\vec{r}\right)\right]}{\partial\rho\left(\vec{r}\right)} - \mu\frac{\partial}{\partial\rho\left(\vec{r}\right)}\left[\int\rho\left(\vec{r}\right)d\vec{r}\right]$$

so we have

$$\frac{\partial A\left[\rho\left(\vec{r}\right)\right]}{\partial\rho\left(\vec{r}\right)} = \frac{\partial E\left[\rho\left(\vec{r}\right)\right]}{\partial\rho\left(\vec{r}\right)} - \mu \tag{14}$$

By replacing this expression in the equation of $\partial A\left[\rho\left(\vec{r}\right)\right]$, we obtained

$$\partial A\left[\rho\left(\vec{r}\right)\right] = \int\left[\frac{\partial E\left[\rho\left(\vec{r}\right)\right]}{\partial\rho\left(\vec{r}\right)} - \mu\right]\partial\rho dr = 0$$

$$\int\frac{\partial E\left[\rho\left(\vec{r}\right)\right]}{\partial\rho\left(\vec{r}\right)}\partial\rho dr = \int\mu\partial\rho dr \tag{15}$$

$$\int\frac{\partial E\left[\rho\left(\vec{r}\right)\right]}{\partial\rho\left(\vec{r}\right)}\partial\rho dr = \mu$$

Using Eq. (9) and calculating the functional derivative of $E\left[\rho\left(\vec{r}\right)\right]$ we obtained

$$\frac{\partial E\left[\rho\left(\vec{r}\right)\right]}{\partial\rho\left(\vec{r}\right)} = \hat{V}_{ext}\left(\vec{r}\right) + \frac{\partial F_{HK}\left[\rho\left(\vec{r}\right)\right]}{\partial\rho\left(\vec{r}\right)} \tag{16}$$

By replacing Eq. (16) in Eq. (15) we obtained

$$\mu = \frac{\partial E\left[\rho\left(\vec{r}\right)\right]}{\partial\rho\left(\vec{r}\right)} = \hat{V}_{ext}\left(\vec{r}\right) + \frac{\partial F_{HK}\left[\rho\left(\vec{r}\right)\right]}{\partial\rho\left(\vec{r}\right)} \tag{17}$$

This equation constitutes the fundamental DFT formalism.

It only remains to determine the expression of the unknown function $F_{HK}\left[\rho\left(\vec{r}\right)\right]$ in Eq. (17), that's why we are brought to use the Kohn-Sham approximations.

2.2 Kohn-Sham (KS) approximations (equations)

This step [10] consists of two approximations to transform the theorems of Hohenberg-Kohn into a practical workable from. However, the real system studied is redefined as a fictitious fermions system without interaction and with the same electronic density $\rho\left(\vec{r}\right)$ characterizing the real system to being up the terms of inter-electronic as corrections of the other terms. Also, single particle orbitals are induced to treat the kinetic energy term of electrons more precisely than under the Thomas-Fermi theory.

2.2.1 First approximation of KS

The transformation of the real interactive system as a fictitious non-interactive system. Considering the first HK theorem, the functional $E\left[\rho\left(\vec{r}\right)\right]$ can be written as indicated in Eq. (9):

$$E\left[\rho\left(\vec{r}\right)\right] = F_{HK}\left[\rho\left(\vec{r}\right)\right] + \int\hat{V}_{ext}\left(\vec{r}\right)\rho\left(\vec{r}\right)d\vec{r} \tag{18}$$

The functional $F_{HK}\left[\rho\left(\vec{r}\right)\right]$ is independent from $\hat{V}_{ext}\left(\vec{r}\right)$ (external potential), and it is valid and applies regardless of the system studied. It contains a component of kinetic energy of electron $\hat{T}_e\left[\rho\left(\vec{r}\right)\right]$, and another component corresponding to the mutual coulomb interaction between electrons $\hat{V}_{e-e}\left[\rho\left(\vec{r}\right)\right]$. The minimization of this functional with the constraint of preservation of the number of particles M:

$\int\rho\left(\vec{r}\right)d\vec{r} = M$gives us directly the total energy of the system and the charge density of ground state from which all other physical properties can be extracted:

$$F_{HK}\left[\rho\left(\vec{r}\right)\right] = T_S\left[\rho\left(\vec{r}\right)\right] + E_H\left[\rho\left(\vec{r}\right)\right] + E_{XC}\left[\rho\left(\vec{r}\right)\right] + V_{ext}\left[\rho\left(\vec{r}\right)\right] \tag{19}$$

Where

$T_S\left[\rho\left(\vec{r}\right)\right]$: is the kinetic energy of non-interactive electron gas.

$E_H\left[\rho\left(\vec{r}\right)\right] = \frac{1}{2}\iint\frac{\rho\left(\vec{r}\right)\rho\left(\vec{r}'\right)}{\left|\vec{r}-\vec{r}'\right|}d\vec{r}d\vec{r}'$: is the chemical coulomb interaction between electrons (Hartree term).

$E_{XC}\left[\rho\left(\vec{r}\right)\right]$: the additional functional describing interactions inter-electronic not obtained from non-interactive system.

$V_{ext}\left[\rho\left(\vec{r}\right)\right] = \int\hat{V}_{ext}\left(\vec{r}\right)\rho\left(\vec{r}\right)d\vec{r}$: is the external potential.

The functional $E_{XC}\left[\rho\left(\vec{r}\right)\right]$ is called "exchange-correlation energy," this term contain all differences between fictitious non-interactive system and the real interactive system.

$$E_{XC}\left[\rho\left(\vec{r}\right)\right] = \left(T\left[\rho\left(\vec{r}\right)\right] - T_S\left[\rho\left(\vec{r}\right)\right]\right) + \left(V_{e-e}\left[\rho\left(\vec{r}\right)\right] - V_H\left[\rho\left(\vec{r}\right)\right]\right) \tag{20}$$

where:

$T\left[\rho\left(\vec{r}\right)\right]$: is the real kinetic energy (of real interactive system).

$T_S\left[\rho\left(\vec{r}\right)\right]$: is the non-interactive fermions system energy of KS.

$T\left[\rho\left(\vec{r}\right)\right] - T_S\left[\rho\left(\vec{r}\right)\right]$: this difference is low and is neglected.

$E_{XC}\left[\rho\left(\vec{r}\right)\right]$ traduced only the difference between Colombian energy of the real system $V_{e-e}\left[\rho\left(\vec{r}\right)\right]$ and the Colombian energy of non-interactive fermions system $V_H\left[\rho\left(\vec{r}\right)\right]$ of KS.

$$E_{XC}\left[\rho\left(\vec{r}\right)\right] \cong V_{e-e}\left[\rho\left(\vec{r}\right)\right] - V_H\left[\rho\left(\vec{r}\right)\right] \tag{21}$$

2.2.2 Second approximation of KS

This second approximation is based on the formulation of kinetic energy using an orbital approach. The exact formulation of kinetic energy T for the ground state systems is given by:

$$T = \sum_{i}^{M} n_{i\langle\varphi_i^*|} \frac{-\hbar^2 \nabla^2 \vec{r}}{2m_e} |\varphi_i\rangle \quad (22)$$

where φ_i are natural spin orbital's and n_i their respective occupation number according to the Pauli principle; $0 \leq n_i \leq 1$.

KS have taken advantage of the fictitious non-interactive fermions system to describe the kinetic energy which is also according to the first theorem of KH as functional of the density:

$$T_s\left[\rho\left(\vec{r}\right)\right] = \sum_{i}^{M} n_{i\langle\varphi_i^*|} \frac{-\hbar^2 \nabla^2 \vec{r}}{2m_e} |\varphi_i\rangle \quad (23)$$

T_s is the non-interactive fermions system energy of KS as function as electron density $\rho\left(\vec{r}\right)$.

3. Kohn-Sham equations

The fundamental equation of DFT is expressed by applying variational KH principal [9] as:

$$\mu = \hat{V}_{eff}\left(\vec{r}\right) + \frac{\partial T_s\left[\rho\left(\vec{r}\right)\right]}{\partial \rho\left(\vec{r}\right)} \quad (24)$$

where $\hat{V}_{eff}\left(\vec{r}\right)$ is formulated as a functional of electronic density.

$$\hat{V}_{eff}\left(\vec{r}\right) = \hat{V}_{eff}\left[\rho\left(\vec{r}\right)\right] = \hat{V}_{ext}\left(\vec{r}\right) + \frac{\partial E_H\left[\rho\left(\vec{r}\right)\right]}{\partial \rho\left(\vec{r}\right)} + \frac{\partial E_{XC}\left[\rho\left(\vec{r}\right)\right]}{\partial \rho\left(\vec{r}\right)}$$

$$\hat{V}_{eff}\left(\vec{r}\right) = \hat{V}_{ext}\left(\vec{r}\right) + \int \frac{\rho\left(\vec{r}'\right)}{\left|\vec{r} - \vec{r}'\right|} d\vec{r}' + \hat{V}_{xc}\left(\vec{r}\right) \quad (25)$$

where $\hat{V}_{xc}\left(\vec{r}\right)$ is the exchange-correlation potential functional derivative of $E_{XC}\left[\rho\left(\vec{r}\right)\right]$ by $\rho\left(\vec{r}\right)$:

$$\hat{V}_{xc}\left(\vec{r}\right) = \frac{\partial E_{XC}\left[\rho\left(\vec{r}\right)\right]}{\partial \rho\left(\vec{r}\right)} \text{ and } \rho\left(\vec{r}\right) = \sum_{i=1}^{M} \left|\varphi_i\left(\vec{r}\right)\right|^2 \quad (26)$$

The Schrödinger equation to resolve in the KS approach is:

$$\left[\frac{-\hbar^2 \nabla^2 \vec{r}}{2m_e} + \hat{V}_{eff}\left(\vec{r}\right)\right] \left|\varphi_i\left(\vec{r}\right)\right\rangle = \varepsilon_i \left|\varphi_i\left(\vec{r}\right)\right\rangle \quad (27)$$

The KS equations (Eqs. (25) and (27)) must be resolved with self-coherent method by starting with an initial electronic density. $\hat{V}_{eff}\left(\vec{r}\right)$ is obtained for which the Schrodinger equation of Kohn-Sham (Eq. 27) is resolved, and a new electronic density is then calculated. From this new electronic density, a new $\hat{V}_{eff}\left(\vec{r}\right)$ is determined. This process is repeated with self-coherent method up to that convergence is reached. The new electronic density obtained must be very close to the previous one corresponding to the criterion of convergence fixed before.

At this step, we only need to find the expression of $E_{XC}\left[\rho\left(\vec{r}\right)\right]$.Some exchange-correlation functional can be considered using different functional approximation families like: Local Density Approximation (LDA) [11], Generalized Gradient Approximation (CGA) [12], Meta Generalized Gradient Approximation (Meta CGA) [13], and Hybrid Functional [14].

Functional Family	Dependence
LDA	$\rho\left(\vec{r}\right)$
GGA	$\left\lvert\nabla\rho\left(\vec{r}\right)\right\rvert, \rho\left(\vec{r}\right)$
Meta-GGA	$\left\lvert\nabla\rho\left(\vec{r}\right)\right\rvert, \nabla^2\rho\left(\vec{r}\right), \rho\left(\vec{r}\right)$
Hybrid	Exact Exchange, $\left\lvert\nabla\rho\left(\vec{r}\right)\right\rvert, \rho\left(\vec{r}\right)$

The order of accuracy is increasing from top to bottom. In the formalism of exchange-correlation functional, $E_{XC}\left[\rho\left(\vec{r}\right)\right]$ is presented like an interaction between the electronic density $\rho\left(\vec{r}\right)$ and density energy depending on $\rho\left(\vec{r}\right)$: $\varepsilon_{XC}\left[\rho\left(\vec{r}\right)\right]$ with

$$E_{XC}\left[\rho\left(\vec{r}\right)\right] = \int \varepsilon_{XC}\left[\rho\left(\vec{r}\right)\right]\rho\left(\vec{r}\right)d\vec{r} \tag{28}$$

$\varepsilon_{XC}\left[\rho\left(\vec{r}\right)\right]$ is considered as a summation of the contribution of exchange and correlation, (with) $\varepsilon_{XC}\left[\rho\left(\vec{r}\right)\right] = \varepsilon_{X}\left[\rho\left(\vec{r}\right)\right] + \varepsilon_{C}\left[\rho\left(\vec{r}\right)\right]$and

$$E_{XC}\left[\rho\left(\vec{r}\right)\right] = E_{X}\left[\rho\left(\vec{r}\right)\right] + E_{C}\left[\rho\left(\vec{r}\right)\right] = \int \varepsilon_{X}\left[\rho\left(\vec{r}\right)\right]\rho\left(\vec{r}\right)d\vec{r} + \int \varepsilon_{C}\left[\rho\left(\vec{r}\right)\right]\rho\left(\vec{r}\right)d\vec{r} \tag{29}$$

3.1 Local Density Approximation (LDA)

In this work we will only use the LDA approximation in our calculations. In this approximation, the electronic density can be treated locally in the form of uniform electrons of gas.

In other words, this approach is to perform the following two hypotheses:

- The exchange-correlation is dominated by the density at the point $\vec{r}$.
- The density $\rho\left(\vec{r}\right)$ is a function slowly varying with $\vec{r}$.

The fundamental hypothesis contained in the LDA formalism is to consider that the contribution of $E_{XC}\left[\rho\left(\vec{r}\right)\right]$ to the total energy of the system can be added by cumulated method of each portion of the non-uniform gas as it was locally uniform.

$$E_{XC}^{LDA}\left[\rho\left(\vec{r}\right)\right] = \int \varepsilon_{XC}^{LDA}\left[\rho\left(\vec{r}\right)\right]\rho\left(\vec{r}\right)d\vec{r} \tag{30}$$

Where $\varepsilon_{XC}^{LDA}\left[\rho\left(\vec{r}\right)\right]$ represent the exchange-correlation energy by electron in a system with electron in mutual correlation with uniform density $\rho\left(\vec{r}\right)$.

Using $\varepsilon_{XC}^{LDA}\left[\rho\left(\vec{r}\right)\right]$, the exchange-correlation potential $V_{XC}^{LDA}\left(\vec{r}\right)$ can be obtained like:

$$V_{XC}^{LDA}\left(\vec{r}\right) = \frac{\partial\left(\rho\left(\vec{r}\right)\varepsilon_{XC}^{LDA}\left[\rho\left(\vec{r}\right)\right]\right)}{\partial\rho\left(\vec{r}\right)} \tag{31}$$

There are several forms for the term of exchange and correlation of a homogeneous electron gas, among others those of Kohn and Sham [10], Wigner [15], Perdew and Wang [16], and Hedin and Lundqvis [17], Ceperly and Alder [18]. The last one is the most used.

3.2 Approximation of Ceperley and Alder

Ceperley and Alder [18] used the exchange energy of uniform electrons gas given by Dirac formula as:

$$E_X\left[\rho\left(\vec{r}\right)\right] = -C_X\int \rho^{\frac{4}{3}}\left(\vec{r}\right)d\vec{r} \tag{32}$$

$$\varepsilon_X^{LDA}\left[\rho\left(\vec{r}\right)\right] = -C_X\int \rho^{\frac{1}{3}}\left(\vec{r}\right)d\vec{r} \tag{33}$$

with $C_X = \frac{3}{4}\left(\frac{3}{\pi}\right)^{\frac{1}{3}}$, $\varepsilon_X^{LDA}\left[\rho\left(\vec{r}\right)\right]$ can be expressed like $\varepsilon_X^{LDA}\left[\rho\left(\vec{r}\right)\right] = -0.738694\rho^{\frac{1}{3}}\left(\vec{r}\right)$.

The correlation energy $\varepsilon_C^{LDA}\left[\rho\left(\vec{r}\right)\right]$ is derived on the second order of Moller-Plesset perturbation theory [19–21]:

$$\varepsilon_C^{LDA}\left[\rho\left(\vec{r}\right)\right] = aLn\left(1+\frac{b}{r_s}+\frac{b}{r_s^2}\right) \tag{34}$$

With $a = \frac{Ln2-1}{2\pi^2} = -0.01556111$ and $b = 20.4562557$.

$r_s$is the Wigner-Seitz density parameter, in atomic unit we have $r_s = \left(\frac{4\pi\rho(\vec{r})}{3}\right)^{-1/3}$.

Considering the correlation term $\varepsilon_C^{LDA}\left[\rho\left(\vec{r}\right)\right]$, no explicit analytical expression is known. Several different parameterizations have been proposed since the early 1970's, and the most accurate results are based on quantum Monte Carlo simulations of Ceperley and Alder [18]. The most approximation commonly used today is that of Zunger [22] and who parameterized the Ceperley and Alder correlation functional for the spin-polarized electron gas and non-spin-polarized electron gas by the equation:

$$\varepsilon_C^{PZ}\left[\rho\left(\vec{r}\right)\right] = \begin{cases} A \times Lnr_s + B + C \times r_s \times Lnr_s + D \times r_s, r_s \leq 1 \\ \gamma/(1 + \beta_1 \times r_s + \beta_2 \times r_s), r_s < 1 \end{cases} \tag{35}$$

For $r_s \leq 1$, $A^U = 0.0311$, $B^U = -0.048$, $C^U = 0.002$ and $D^U = -0.0116$.
For $r_s < 1$, $\gamma = -0.1423$, $\beta_1 = 1.0529$, and $\beta_2 = 0.3334$.
The improvements of the LDA approach must consider the gas of electrons in its real form (non-uniform and non-local), GGA, meta-GGA, and hybrid functional allow gradually approach to make in consideration their two effects.

4. Resolution of the KS Schrödinger equation

Combining between Eqs. (25) and (27) and using the expression $\rho\left(\vec{r}\right) = \sum_{i=1}^{M}\left|\varphi_i\left(\vec{r}\right)\right|^2$, where M is the number of electrons, Eq. (27) becomes:

$$\left[\frac{-\hbar^2\nabla^2\vec{r}}{2m_e} + \hat{V}_{ext}\left(\vec{r}\right) + \int\frac{\rho\left(\vec{r}'\right)}{\left|\vec{r} - \vec{r}'\right|}d\vec{r}' + \hat{V}_{xc}\left(\vec{r}\right)\right]\left|\varphi_i\left(\vec{r}\right)\right\rangle = \varepsilon_i\left|\varphi_i\left(\vec{r}\right)\right\rangle \tag{36}$$

The only unknown value in this expression is $\hat{V}_{xc}\left(\vec{r}\right)$ (exchange-correlation potential).

Using LDA approximation of $\hat{V}_{xc}\left(\vec{r}\right)$ (Eq. (30)),Where $\varepsilon_{xc}^{LDA}\left[\rho\left(\vec{r}\right)\right] = \varepsilon_x^{LDA}\left[\rho\left(\vec{r}\right)\right] + \varepsilon_c^{LDA}\left[\rho\left(\vec{r}\right)\right]$ with

$\varepsilon_{xc}^{LDA}\left[\rho\left(\vec{r}\right)\right]$: The exchange-correlation energy of one electron in a system of electron on mutual interaction with electronic density $\rho\left(\vec{r}\right)$.

$\varepsilon_x^{LDA}\left[\rho\left(\vec{r}\right)\right]$: The exchange energy of one electron.

$\varepsilon_c^{LDA}\left[\rho\left(\vec{r}\right)\right]$: The correlation energy of one electron.

Eq. (36) becomes:

$$\left[\frac{-\hbar^2\nabla^2\vec{r}}{2m_e} + \hat{V}_{ext}\left(\vec{r}\right) + \int\frac{\rho\left(\vec{r}'\right)}{\left|\vec{r} - \vec{r}'\right|}d\vec{r}' + \frac{\partial\left(\rho\left(\vec{r}\right)\varepsilon_{xc}^{LDA}\left[\rho\left(\vec{r}\right)\right]\right)}{\partial\rho\left(\vec{r}\right)}\right]\left|\varphi_i\left(\vec{r}\right)\right\rangle = \varepsilon_i\left|\varphi_i\left(\vec{r}\right)\right\rangle \tag{37}$$

$$\left[\frac{-\hbar^2\nabla^2\vec{r}}{2m_e}+\hat{V}_{ext}\left(\vec{r}\right)+\int\frac{\rho\left(\vec{r}'\right)}{\left|\vec{r}-\vec{r}'\right|}d\vec{r}'+\frac{\partial\left(\rho\left(\vec{r}\right)\left(\varepsilon_x^{LDA}\left[\rho\left(\vec{r}\right)\right]+\varepsilon_c^{LDA}\left[\rho\left(\vec{r}\right)\right]\right)\right)}{\partial\rho\left(\vec{r}\right)}\right]\left|\varphi_i\left(\vec{r}\right)\right\rangle = \varepsilon_i\left|\varphi_i\left(\vec{r}\right)\right\rangle \tag{38}$$

$$\left[\frac{-\hbar^2\nabla^2\vec{r}}{2m_e}+\hat{V}_{ext}\left(\vec{r}\right)+\int\frac{\rho\left(\vec{r}'\right)}{\left|\vec{r}-\vec{r}'\right|}d\vec{r}'+\frac{\partial\left(\rho\left(\vec{r}\right)\left(\varepsilon_x^{LDA}\left[\rho\left(\vec{r}\right)\right]\right)\right)}{\partial\rho\left(\vec{r}\right)}+\frac{\partial\left(\rho\left(\vec{r}\right)\left(\varepsilon_c^{LDA}\left[\rho\left(\vec{r}\right)\right]\right)\right)}{\partial\rho\left(\vec{r}\right)}\right]\left|\varphi_i\left(\vec{r}\right)\right\rangle = \varepsilon_i\left|\varphi_i\left(\vec{r}\right)\right\rangle \tag{39}$$

where

$$\hat{H}_{KS} = -\frac{1}{2}\nabla^2\vec{r}+\hat{V}_{ext}\left(\vec{r}\right)+\int\frac{\rho\left(\vec{r}'\right)}{\left|\vec{r}-\vec{r}'\right|}d\vec{r}'+\frac{\partial\left(\rho\left(\vec{r}\right)\left(\varepsilon_x^{LDA}\left[\rho\left(\vec{r}\right)\right]\right)\right)}{\partial\rho\left(\vec{r}\right)}+\frac{\partial\left(\rho\left(\vec{r}\right)\left(\varepsilon_c^{LDA}\left[\rho\left(\vec{r}\right)\right]\right)\right)}{\partial\rho\left(\vec{r}\right)} \tag{40}$$

$$\hat{V}_{eff}\left(\vec{r}\right) = \hat{V}_{ext}\left(\vec{r}\right)+\int\frac{\rho\left(\vec{r}'\right)}{\left|\vec{r}-\vec{r}'\right|}d\vec{r}'+\frac{\partial\left(\rho\left(\vec{r}\right)\left(\varepsilon_x^{LDA}\left[\rho\left(\vec{r}\right)\right]\right)\right)}{\partial\rho\left(\vec{r}\right)}+\frac{\partial\left(\rho\left(\vec{r}\right)\left(\varepsilon_c^{LDA}\left[\rho\left(\vec{r}\right)\right]\right)\right)}{\partial\rho\left(\vec{r}\right)} \tag{41}$$

In the LDA approximation, Eq. (38) must be the departure of our programming calculations. All terms in Eq. (38) are known and remain to define $\hat{V}_{ext}\left(\vec{r}\right)$ and basis of functions $\varphi_i\left(\vec{r}\right)$.

$\hat{V}_{ext}\left(\vec{r}\right)$ and $\varphi_i\left(\vec{r}\right)$ are selected as the case we want to treat. We will give further their expressions.

Eq. (37) must be solved self-consistently by starting at a certain density. An effective potential $\hat{V}_{eff}\left(\vec{r}\right)$ is obtained for which the Schrodinger equation of Kohn-Sham is resolved, and a new electronic density is determined. This process is repeated self-consistently until convergence is reached, and the new electronic density obtained must be very close to the previous one.

4.1 Self-coherent technical to resolve KS Schrödinger equation

The idea is that do not directly resolve Eq. (39), but to write previously the $\phi_m\left(\vec{r}\right)$ in a finite basis function $\phi_p^b\left(\vec{r}\right)$ as:

$$\phi_m\left(\vec{r}\right) = \sum_{p=1}^{p} C_p^m \phi_p^b\left(\vec{r}\right) \tag{42}$$

when $m = n.\vec{k}$, $\vec{k}$: is the wave vector belonging the first Brillouin zone in the case of crystal lattice.

The resolution of Eq. (38) consists to determine the coefficients C_p^m necessary to express $\phi_m\left(\vec{r}\right)$ in a given basis $\phi_p^b\left(\vec{r}\right)$.

We need to search a basis to get as close as possible to $\phi_m\left(\vec{r}\right)$ with p having a finite value.

Eq. (37) becomes:

$$\begin{bmatrix} \cdots & & \cdots \\ \cdots & \left\langle\phi_i^b\middle|\hat{H}_{KS}\middle|\phi_j^b\right\rangle - \varepsilon_m\left\langle\phi_i^b\middle|\phi_j^b\right\rangle & \cdots \\ \cdots & & \cdots \end{bmatrix} \begin{bmatrix} C_1^m \\ \cdots \\ C_p^m \end{bmatrix} = \begin{bmatrix} 0 \\ \cdots \\ 0 \end{bmatrix}, \tag{43}$$

in which one can identify the matrix Elements of Hamiltonian of single particle and the elements of the recovery matrix, $\left(H_{ij} - \varepsilon_m S_{ij}\right)C_p^m = 0$, where $\hat{H}_{ij} = \left\langle\phi_i^b\middle|\hat{H}_{KS}\middle|\phi_j^b\right\rangle$ and $S_{ij} = \left\langle\phi_i^b\middle|\phi_j^b\right\rangle$ respectively represent the Hamiltonian matrix and the recovery matrix.

For a solid, these equations need to be resolved for each vector k in the Brillouin zone. This secular equation system is linear with the energy. This system constitutes a problem of determination of the proper values ε_m and proper functions $\phi_i^k\left(\vec{r}\right)$ that much we know within the Hartree theory and is commonly solved from standard numerical methods. The diagonalization of the Hamiltonian matrix provides p proper values clean and p sets values of coefficients that express each p proper functions in a given basis.

More p is big, more the proper functions are precise, but the matrix diagonalization time is also particularly high.

4.2 Self-consistent cycle

Eq. (39) must be resolved in an iterative way in self-consistent cycle procedure. The procedure starts by the definition of a density of departure corresponding to a determined geometry core. Generally, the initial density is constituted from a superposition of atomic densities: $\rho^{in} = \rho^{crystal} = \sum_{at}\rho^{at}$.

When the elements of the Hamiltonian matrix and recovery matrix were calculated, proper vectors and proper functions are determined from the diagonalization of the matrix $(H_{ij} - \varepsilon_m S_{ij})C_p^m = 0$. Following the principle of Aufbau, the orbitals are filled, a new density is determined:

$$\boxed{\rho^{out}\left(\vec{r}\right) = \sum_{out}\left|\phi_i\left(\vec{r}\right)\right|^2},$$

This step concludes the first cycle. At this stage of the process, acceleration of convergence is generally used to generate a new density realized from a mixture between this output density $\rho^{out}(r)$ and density of entry of this cycle and there $\rho^{in}(r)$. One of the simplest procedures concerning this mixture can be formulated as: $\rho_{i+1}^{in}(r) = (1-\alpha)\rho_i^{in}(r) + \alpha\rho_i^{out}(r)$ with $0 < \alpha < 1$. α is the mixture parameter an i corresponds to iterative cycle number. Density of the new entry $\rho_{i+1}^{in}(r)$ is then introduced into a second self-consistent cycle. This process is repeated for an iterative manner until it has the convergence criterion (difference between $\rho^{out}(r)$ and $\rho^{in}(r)$) initially fixed is reached. When convergence is reached, the energy of the ground state of the system is reached (**Figure 1**).

Once we obtain $\rho^{out}\left(\vec{r}\right)$ for the first cycle, we inject it like a new value of $\rho^{in}\left(\vec{r}\right)$ and we repeat this iterative operation until it is that the fixed precision at the beginning is reached.

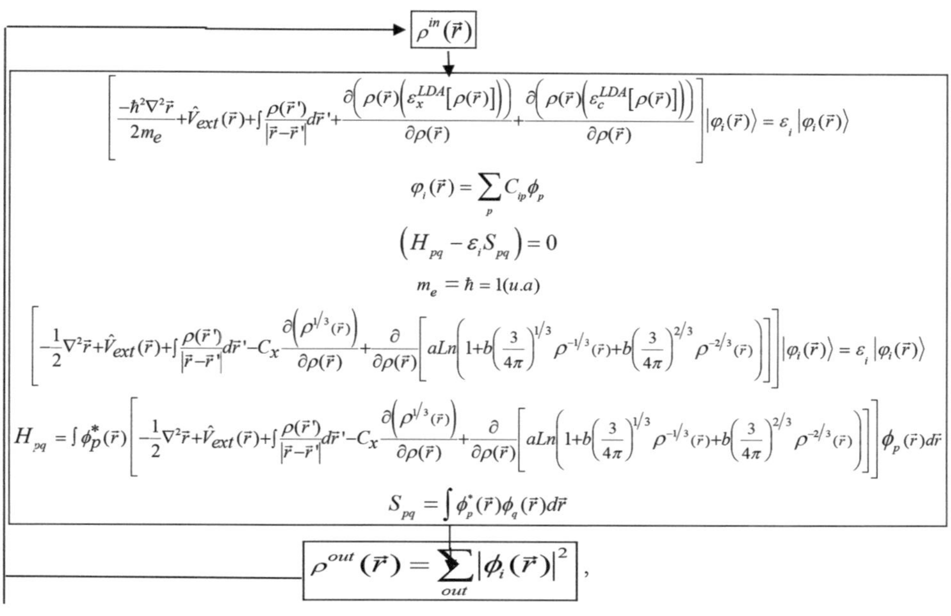

Figure 1.
Schematic representation of the self-consistent cycle.

Precision = $\rho^{out}\left(\vec{r}\right) - \rho^{in}\left(\vec{r}\right)$

$$\rho^{in}\left(\vec{r}\right) = \rho^{orbital}\left(\vec{r}\right) = \sum_{at} \rho^{at}\left(\vec{r}\right),$$

We can use $\phi_k^n\left(\vec{r}\right) = e^{i\vec{k}.\vec{r}} u_k^n\left(\vec{r}\right)$, where $e^{i\vec{k}.\vec{r}}$ is a plane wave and $u_k^n\left(\vec{r}\right)$ a function processing Lattice periodicity (Bloch theorem $u_k^n\left(\vec{r} + \vec{k}\right) = u_k^n\left(\vec{r}\right)$).

5. Time dependent density functional theory (TDDFT) calculations

The Eigen values of Kohn-Sham does not match corresponding to the required energy for excited electrons in a correlated electron system with N electrons, because the Kohn-Sham [10] approach replaces the interacting electron problem by an independent electron problem. There is a way to study the potential and the wave function to determine the excitation energies. It is the use of the TDDFT theory which takes into account the time parameter. The excitation energy of a system can then be obtained from the response function to extreme electron density perturbations.

5.1 Rung-Gross theorems

TDDFT is based on the Rung-Gross theorem [23, 24] that is the analog of the time-dependent of Hohenberg-Kohn theorems [9]. The theorems are cited below:

5.1.1 First theorem

The time depends on the external potential $\hat{V}_{ext}\left(\vec{r}, t\right)$ is determined by the time-dependent electron density $\rho\left(\vec{r}, t\right)$ to a nearly additive function for an initial state ψ a state$\psi(t = 0)$.

5.1.2 Second theorem

By difference with the ground state, the variational principle, which stated that there is a minimum associated with the total energy does not exist for the time-dependent systems because the energy is not a conserved quantity. A similar amount of energy, which is applied on the stationary principle, is defined as:

$$\mathrm{A}[\psi] = \int_{t_0}^{t_1} dt \langle \psi(t) | i \frac{\partial}{\partial t} - \hat{H}(t) | \psi(t) \rangle \tag{44}$$

where A is called the action, and $\psi(t)$ is a function of the time-dependent poly electronic wave function.

5.2 The time-dependent approach of Kohn-Sham

The time-dependent approach [10] of Kohn-Sham shows an equation with partial derivatives for the effective system such as

$$i\frac{\partial}{\partial t}\varphi_i\left(\vec{r},t\right) = H\varphi_i\left(\vec{r},t\right) = \left[-\frac{1}{2}\nabla^2 + \hat{V}_{eff}\left(\rho\left(\vec{r},t\right)\right)\right]\varphi_i\left(\vec{r},t\right) \tag{45}$$

$$\rho\left(\vec{r},t\right) = \sum_{i=1}^{N}\left|\varphi_i\left(\vec{r},t\right)\right|^2 \tag{46}$$

With $\varphi_i\left(\vec{r},t\right)$ is a single electron time-dependent wave function of the Kohn-Sham. $\hat{V}_{eff}\left(\rho\left(\vec{r},t\right)\right)$ defined in Eq. (46) contains in addition to the external field, a Hartree potential, and the exchange-correlation potential such as:

$$\hat{V}_{eff}\left(\rho\left(\vec{r},t\right)\right) = \hat{V}_{ext}\left(\vec{r},t\right) + \frac{\partial E_H\left(\vec{r},t\right)}{\partial\rho\left(\vec{r},t\right)} + \hat{V}_{xc}\left(\rho\left(\vec{r},t\right)\right) \tag{47}$$

The exchange-correlation term $\hat{V}_{xc}\left(\rho\left(\vec{r},t\right)\right)$ is unknown. There is a major difficulty to evaluate it because it does not depend only on the density $\rho\left(\vec{r},t\right)$, but also depends on the prior density at time t in every point in space. The evolution of $\hat{V}_{xc}\left(\rho\left(\vec{r},t\right)\right)$ therefore requires the introduction of the single fundamental approximation of TDDFT.

6. Linear response theory

The TDDFT approach allows one hand to study the perturbation of the system at time t_0, secondly to propagate this disturbance for a time $t > t_0$. The study of the evolution in the propagation of this disturbance leads to the production of the absorption spectrum. This theory [25, 26] is used in the space of frequencies ω rather than in the temporal space. The operation is performed by means of the Fourier transform.

Considering, the continuous function f of the time variable. The Fourier transform of f gives:

$$TF(f) : \omega \mapsto f(\omega) = \frac{1}{\sqrt{2\pi}}\int_{-\infty}^{+\infty} f(t)e^{-i\omega t}dt \tag{48}$$

Considering a system in the ground state of electron density ρ_0 which is subject at $t = t_0$ to a low disturbance of the external potential $\partial\hat{V}_{ext}$ the electron density obtained following this disturbance is, at first order given by:

$$\rho\left(\vec{r},\omega\right) = \rho_0\left(\vec{r}\right) + \delta\rho\left(\vec{r},\omega\right) \tag{49}$$

In the interacting electrons system, the external potential is also written as the sum of the external potential calculated in the ground state, supplemented with a disruptive potential:

$$\hat{V}_{ext}\left(\vec{r},\omega\right) = \hat{V}_{ext}\left(\vec{r}\right) + \delta\hat{V}_{ext}\left(\vec{r},\omega\right) \tag{50}$$

The perturbation $\delta\rho$ depends only on the potential δV_{ext} and takes the expression:

$$\delta\rho\left(\vec{r},\omega\right) = \int_{r'} \chi\left(\vec{r},\vec{r}',\omega\right)\delta\hat{V}_{ext}\left(\vec{r}',\omega\right)d\vec{r}' \tag{51}$$

$\chi\left(\vec{r},\vec{r}',\omega\right)$: represents the response function of the non-interacting electrons system. The evolution of χ is complex, but it can be simplified by using the Kohn-Sham approach that redefines $\delta\rho\left(\vec{r},\omega\right)$ according to the equation:

$$\delta\rho\left(\vec{r},\omega\right) = \int_{r'} \chi_{KS}\left(\vec{r},\vec{r}',\omega\right)\delta\hat{V}_{eff}\left(\vec{r}',\omega\right)d\vec{r}' \tag{52}$$

where $\chi_{KS}\left(\vec{r},\vec{r}',\omega\right)$, according to Kohn-Sham response is easily calculated using the equation:

$$\chi_{KS}\left(\vec{r},\vec{r}',\omega\right) = \lim_{\eta\to 0^+} \sum_{p=1}^{\infty}\sum_{q=1}^{\infty}\left(f_p - f_q\right)\frac{\varphi_p^*\left(\vec{r}\right)\varphi_q\left(\vec{r}\right)\varphi_q^*\left(\vec{r}'\right)\varphi_p\left(\vec{r}'\right)}{\omega - \omega_{p\to q} + i\eta} \tag{53}$$

This expression (52) represents the various excitations between the occupied Kohn-Sham orbital's φ_p and unoccupied φ_q, for a total number N of orbital's approaching infinity. The occupation of the orbital's number p and q are respectively denoted f_p and f_q. The frequency $\omega_{p\to q} = \varepsilon_p - \varepsilon_q$; where ε_p and ε_q are the eigenvalues respectively associated with wave functions φ_p and φ_q. The infinitesimal positive number η is introduced to reflect the causality of the system response.

Disruption of the effective potential $\delta\hat{V}_{eff}$ (Kohn-Sham potential) can be written as the sum of three terms:

- the Coulomb term: $\int_{r'} \frac{\delta\rho\left(\vec{r}',\omega\right)}{\left|\vec{r}-\vec{r}'\right|}d\vec{r}'$
- the external term: $\delta\hat{V}_{ext}\left(\vec{r},\omega\right)$
- the exchange-correlation term: $\delta\hat{V}_{xc}\left(\vec{r},\omega\right)$

Such:

$$\delta\hat{V}_{eff}\left(\vec{r},\omega\right) = \delta\hat{V}_{ext}\left(\vec{r},\omega\right) + \int_{r'} \frac{\delta\rho\left(\vec{r}',\omega\right)}{\left|\vec{r}-\vec{r}'\right|}d\vec{r}' + \delta\hat{V}_{xc}\left(\vec{r},\omega\right) \tag{54}$$

Following a Fourier transformation, it is customary to write disturbance of the exchange-correlation potential as:

$$\delta\hat{V}_{xc}\left(\vec{r},\omega\right) = TF\left(\int_{r'}\int_{t'} f_{xc}(r,t,r',t')\delta\rho(r',t')dt'dr'\right) \tag{55}$$

Where

$$f_{xc}(r,t,r',t') = \frac{\delta\hat{V}_{xc}\left(\vec{r},\omega\right)}{\delta\rho(r',t')}, \tag{56}$$

$f_{xc}(r,t,r',t')$ is the expression of exchange and correlation core.

In this step of the calculation, the development of the disturbance of the electron density (Eq. 51) through the expression of the effective potential disturbance (Eq. 52) gives:

$$\delta\rho\left(\vec{r},\omega\right) = \int_{r'}\chi_{KS}\left(\vec{r},\vec{r}',\omega\right)\delta\hat{V}_{ext}\left(\vec{r}',\omega\right)d\vec{r}' + \int_{r_1}\int_{r_2}\chi_{KS}\left(\vec{r},\vec{r}_1,\omega\right)f_{Hxc}\left(\vec{r}_1,\vec{r}_2,\omega\right)\delta\rho\left(\vec{r}_2,\omega\right)dr_2dr_1 \tag{57}$$

Where

$$f_{Hxc}\left(\vec{r},\vec{r}',\omega\right) = \frac{1}{\left|\vec{r}-\vec{r}'\right|} + f_{xc}\left(\vec{r},\vec{r}',\omega\right) \tag{58}$$

$f_{Hxc}\left(\vec{r},\vec{r}',\omega\right)$ is the expression of the Hartree exchange-correlation core.

Expression of the disturbance of the electron density is known both in the interacting electron system (Eq. 57) and non-interacting electrons system (Eq. 51). The equality of these two expressions leads to that of the response function of the interacting electron system:

$$\chi\left(\vec{r},\vec{r}',\omega\right) = \chi_{KS}\left(\vec{r},\vec{r}',\omega\right) + \int_{r_1}\int_{r_2}\chi\left(\vec{r},\vec{r}_1,\omega\right)f_{Hxc}\left(\vec{r}_1,\vec{r}_2,\omega\right)\chi_{KS}\left(\vec{r}_2,\vec{r}',\omega\right)dr_2dr_1 \tag{59}$$

Equation of the susceptibility χ is important because this linear response has poles exactly to the energies of the electronic transitions of the system.

- When $f_{Hxc}\left(\vec{r}_1,\vec{r}_2,\omega\right) = 0$, the electronic transitions are exactly equal to those given by the Kohn-Sham system in the ground state. Their oscillator strengths are then given by the poles of χ_{KS}.
- When $f_{Hxc}\left(\vec{r}_1,\vec{r}_2,\omega\right) \neq 0$, the excitation energies are corrected by the exchange-correlation Hartree core, and the oscillator strengths of the transitions are given by the poles χ.

7. Resolution of time-dependent problem

7.1 Adiabatic approximation

No analytical expression exists for the time-dependent exchange correlation potential. So, we work using the adiabatic approximation [26] by ignoring the dependence on previous electron densities of these two terms and use only the instantaneous electron density. Furthermore, when the time-dependent potential changes slowly, thus adiabatically, the system remains in its instantaneous ground state. The exchange-correlation potential developed in the DFT to describe the ground state of the system can then be translated to time-dependent systems. The potential and the adiabatic exchange-correlation core then take local forms in time given to equations (Eq. 60) and (Eq. 59). It is possible to note that in this approximation, the exchange-correlation core is independent of the frequency of the disturbance potential applied to the system.

$$\hat{V}_{xc}^{adia}\left(\vec{r},t\right) = \hat{V}_{xc}^{DFT}\left[\rho\left(\vec{r}\right)\right]_{\rho\left(\vec{r}\right)=\rho\left(\vec{r},t\right)} \tag{60}$$

$$f_{xc}^{adia}\left(\vec{r},t,r',t'\right) = \frac{\delta\hat{V}_{xc}\left(\vec{r}\right)}{\delta\rho\left(\vec{r}'\right)}\delta(t-t') \tag{61}$$

The simplicity of this approximation allows to make after Fourier transform, the exchange-correlation core f_{xc}^{adia} independent of the frequency ω. Therefore, f_{Hxc} also becomes independent of the frequency as:

$$\hat{f}_{Hxc}^{adia}\left(\vec{r},\vec{r}',\omega\right) = f_{Hxc}^{adia}\left(\vec{r},\vec{r}'\right) = \frac{1}{\left|\vec{r}-\vec{r}'\right|} + f_{xc}^{adia}\left(\vec{r},\vec{r}'\right) \tag{62}$$

In the adiabatic approximation, using functional developed for DFT retains weaknesses of these functional. For example, functional LDA and most functional GGA falsely represent the decay of exchange-correlation potential in neutral finite systems.

7.2 Equation with Eigenvalues

When the exchange-correlation core was independent of the frequency (in the adiabatic approximation case), then search for the poles of the response factor by solving a system of equations to the eigenvalues [22, 26] of a matrix equation as the specific form:

$$\sum_{p'=1}^{\infty}\sum_{q'=1}^{\infty} M_{pq,p'q'}\left[f_{Hxc}^{adia}\right]X_{p'q'} = \Omega_{pq}^{2}X_{pq} \tag{63}$$

where p, p' et q, q' respectively represent indices of non-occupied and occupied orbitals for the orbital's base with size tends to infinity. $M_{pq,p'q'}$ is an element of the operator matrix writing such that:

$$M_{pq,p'q'}\left[f_{Hxc}^{adia}\right] = 2W_{pq,p'q'}\left[f_{Hxc}^{adia}\right]\sqrt{\omega_{p\to q}\omega_{p'\to q'}} + \omega_{p\to q}^{2}\delta(pq)(p'q'), \tag{64}$$

The matrix element $\omega_{pq,p'q'}$ is function as the Hartree exchange-correlation core, itself-independent of the frequency as:

$$\omega_{pq,p'q'}\left[f_{Hxc}^{abia}\right] = \int_{r}\int_{r'}\varphi_{p}^{*}\left(\vec{r}\right)\varphi_{p'}^{*}\left(\vec{r}'\right)f_{Hxc}^{adia}\left(\vec{r},\vec{r}'\right)\varphi_{q}\left(\vec{r}\right)\varphi_{q'}\left(\vec{r}'\right)d\vec{r}d\vec{r}' \tag{65}$$

Where X_{pq} et Ω_{pq} are respectively the vectors and eigenvalues.

8. Conclusion

Our main objective is the determination of the electronic structure, so we have proposed the model of the Functional Density Theory dependent and independent of time which is widely used; it has the advantage of taking into account electron correlation directly in its formalism. To test the relevance of these theoretical calculations and the limit of validity of all these approximations used during the development of this theoretical model, a computer simulation model is necessary. Indeed, it will allow us to obtain the eigenvalues and eigenvectors. They provide enough information to access to the electronic structure and total energy. This theoretical investigation may allow to work toward the realistic modeling of the electronic structure of spherical and cylindrical nanoparticles, to provide effective potentials and orbitals that we can employ to calculate photoemission spectra, and may allow to improve the modeling of the ground-state electronic structure of metal surfaces within the framework of the streaked photoemission calculations, and to develop a theoretical framework for dielectric response of surfaces and nanostructures to external IR fields in order to model plasmonic electric fields [27–30] enhancement near nanostructures systems.

Acknowledgements

We are grateful to Professor Uwe Thumm who hosted us for 3 months in his James R. Macdonald laboratory at the Kansas State University in the USA, and who offered us an opportunity to collaborate on this subject, as part of the Fulbright Grant Merit Award.

Author details

Brahim Ait Hammou[1], Abdelhamid El Kaaouachi[1]*, El Hassan Mounir[2], Hamza Mabchour[2], Abdellatif El Oujdi[2], Adil Echchelh[2], Said Dlimi[3] and Driss Ennajih[2]

1 Faculty of Sciences of Agadir, Materials and Physicochemistry of the Atmosphere and Climate Group, Agadir, Morocco

2 Faculty of Sciences Ibn Tofail, Laboratory of Energetic Engineering and Materials, Kenitra, Morocco

3 Faculty of Sciences, Laboratory of Electronics, Instrumentation and Energetic, Physics Department, Chouaib Doukkali, El Jadida, Morocco

*Address all correspondence to: kaaouachi21@yahoo.fr

References

[1] Ambrosio MJ, Thumm U. Energy-resolved attosecond interferometric photoemission from Ag(111) and Au (111) surfaces. Physical Review A. 2018; **97**:043431. DOI: 10.1103/PhysRevA.97.043431

[2] Ambrosio MJ, Thumm U. Electronic structure effects in spatiotemporally resolved photoemission interferograms of copper surfaces. Physical Review A. 2017;**96**:051403(R)

[3] Leone SR, McCurdy CW, Burgdörfer J, Cederbaum LS, Chang Z, Dudovich N, et al. What will it take to observe processes in 'real time'? Nat. Photon. 2014;**8**:162-166. DOI: 10.1038/nphoton.2014.48

[4] Calegari F, Sansone G, Stagira S, Vozzi C, Nisoli M. Advances in attosecond science. J. Phys. B. 2016;**49**: 062001. DOI: 10.1088/0953-4075/49/6/062001

[5] Goulielmakis E, Schultze M, Hofstetter M, Yakovlev VS, Gagnon J, Uiberacker M, et al. Single-cycle nonlinear optics. Science. 2008;**320**: 1614-1617. DOI: 10.1126/science.1157846

[6] Ambrosio MJ, Thumm U. Comparative time-resolved photoemission from Cu(100) and Cu (111) surfaces. Surfaces Phys. Rev. A. 2016;**94**:063424

[7] Tanuma S, Powell CJ, Penn DR. Calculations of electron inelastic mean free paths. IX. Data for 41 elemental solids over the 50 eV to 30 keV range. Surface and Interface Analysis. 2011;**43**: 689-713. DOI: 10.1002/sia.3522

[8] Roth F, Lupulescu C, Darlatt E, Gottwald A, Eberhardt W. Angle resolved photoemission from Cu single crystals: Known facts and a few surprises about the photoemission process. J. Electron Spectrosc. 2016;**208**:2-10. DOI: 10.1016/j.elspec.2015.09.006

[9] Hohenberg P, Kohn W. Inhomogeneous electron gas. Physical Review B. 1969;**136**:864. DOI: 10.1103/PhysRev.136.B864

[10] Kohn W, Sham LJ. Self-consistent equations including exchange and correlation effects. Physical Review A. 1965;**140**:1133. DOI: 10.1103/PhysRev.140.A1133

[11] Sahni V, Bohnen KP, Harbola MK. Analysis of the local-density approximation of density-functional theory. Physical Review A. 1988;**37**:1895

[12] Ziesche P, Kurth S, Perdew JP. Density functionals from LDA to GGA. Journal of Computational Materials Science. 1998;**11**(2):122-127

[13] Womack JC, Mardirossian N, Head-Gordon M. Self-consistent implementation of meta-GGA functionals for the ONETEP linear-scaling electronic structure package. The Journal of Chemical Physics. 2016;**145**(20):204114

[14] Koller D, Blaha P, Wien TU, Tran F. Hybrid functionals for solids with an optimized Hartree-Fock mixing parameter. Journal of Physics Condensed Matter. 2013;**25**(43):435503. DOI: 10.1088/0953-8984/25/43/435503

[15] Wigner E. On the interaction of electrons in metals. Physics Review. 1934;**46**:1001. DOI: 10.1103/PhysRev.46.1002

[16] Perdew JP, Wang Y. Accurate and simple analytic representation of the

electron-gas correlation energy. Physical Review B. 1992;**45**:13244. DOI: 10.1103/PhysRevB.45.13244

[17] Hedin L, Lundqvist B. Explicit local exchange-correlation potentials. Journal of Physics C. 1971;**4**:2064. DOI: 10.1088/0022-3719/4/14/022

[18] Ceperley DM, Alder BJ. Ground state of the electron gas by a stochastic method. Physical Review Letters. 1980;**45**:566. DOI: 10.1103/PhysRevLett.45.566

[19] Loos PF, Gill PMW. Leading-order behavior of the correlation energy in the uniform electron gas. International Journal of Quantum Chemistry. 2012;**112**:1712

[20] Moller C, Plesset M. Note on an approximation treatment for many-electron systems. Physics Review. 1934;**46**:618. DOI: 10.1103/PhysRev.46.618

[21] Handler GS. On the removal of the exchange singularity in extended systems. International Journal of Quantum Chemistry. 1988;**33**:173

[22] Zunger A, Ihm J, Cohen ML. Momentum-space formalism for the total energy of solids. Journal of Physics C. 1979;**12**(21):4409. DOI: 10.1088/0022-3719/12/21/009

[23] Runge E, Gross EKU. Density-functional theory for time-dependent systems. Physical Review Letters. 1984;**52**:997. DOI: 10.1103/PhysRevLett.52.997

[24] Schirmer J. Reexamination of the Runge-Gross action-integral functional. Physical Review A. 2012;**86**:012514. DOI: 10.1103/PhysRevA.86.012514

[25] Mermin ND. Contribution à la modélisation électro-thermique de la cellule de commutation MOSFET-Diode. Physique des solides. France: Universitaire, Institut des Sciences Appliquées de Lyon; 2003

[26] Payne MC, Teter MP, Joannopoulos JD. Iterative minimization techniques for *ab initio* total-energy calculations: Molecular dynamics and conjugate gradients. Reviews of Modern Physics. 1992;**64**(4):1045. DOI: 10.1103/RevModPhys.64.1045

[27] Hendrik JM, James D. Special points for Brillouin-zone integrations. Pack Physical Review B. 1976;**13**:5188. DOI: 10.1103/PhysRevB.13.5188

[28] Chevreuil P-A, Brunner F, Thumm U, Keller U, Gallmann L. Breakdown of the single-collision condition for soft x-ray high harmonic generation in noble gases. Optica. 2022. DOI: 10.1364/OPTICA.471084

[29] Saydanzad E, Li J, Thumm U. Strong-field ionization of plasmonic nanoparticles. Physical Review A. 2022. DOI: 10.1103/PhysRevA.106.033103

[30] Thumm U. Enhanced extreme ultraviolet high-harmonic generation from chromium- doped magnesium oxide. Applied Physics Letters. 2021. DOI: 10.1063/5.0047421

Chapter 3

Distinct Roles of the Principal Exchange-Correlation Energy and the Secondary Correlation Energy Functionals in the MGC-SDFT-UHFD Decoupling

Masami Kusunoki

Abstract

The Kohn-Sham formalism for the density functional theory (DFT) proposed a half-century ago has been the extensive motive force for the material science community, despite it is incomplete because of its problematic notion of eternally-unknown correlation energy functional including a separated part of kinetic energy. Here, we widely explain an alternative method recently discovered by us, i.e. the multiple grand canonical spin DFT (MGC-SDFT) in the unrestricted Hartree-Fock-Dirac (MGC-SDFT-UHFD) approximation. It is proved that the correlation energy functional consists of well-defined principal and secondary parts: the former yields the principal internal energy functional responsible for a set of the one-body quasi-particle spectra defined by the respective ground and excited states with each natural LCAO-MO as well as a set of the expected values of Heisenberg spin Hamiltonian, and the latter does a well-defined spin-dependent perturbation energy responsible for some many-body effects. An application will be made to explain why the water-splitting S_1-state Mn_4CaO_5-clusters in photosystem II can exhibit two different EPR signals, called "g4.8" and "g12-multiline". Moreover, the *secondary correlation energy part* will be shown to promote Cooper-pairings of Bloch-electrons near Fermi level in the superconductor, provided that their eigenstates might be exactly determined by the MGC-SDFT-UHFD method.

Keywords: spin density functional theory (SDFT), LCAO-natural molecular orbitals (NMO), principal exchange-correlation energy, Heisenberg spin Hamiltonian, secondary correlation energy, superconductivity

1. Introduction

In this chapter, we aim to explain why the predominant Kohn-Sham formalism of density functional theory (KS-DFT) based on the variational principle with respect to the electron density in a closed N-electron system [1–6], must be stated as incomplete, during a number of active works motivated on it (e.g [7–12]) still continuing, by pushing out the alternative electron density functional theory based on the multiple grand canonical quantum statistical variational principle capable of generating a large enough number of quantized energy levels of the ground and excited states in the unrestricted Hartree-Fock-Dirac approximation taking account of the explicit principal exchange-correlation energy functional -K_{XC}. This ultimate theory has been recently developed and called "multiple grand canonical spin DFT in the UHFD approximation (MGC-SDFT-UHFD method)" [13]. Moreover, we aim to present here a compact text of this ultimate MGC-SDFT-UHFD method in Sections 2.1 and 2.2 in order to help not only the reader's understanding but also some program developers to challenge this painful-but-promising project to revise some codes associated with this paradigm shift from KS-DFT to MGC-SDFT-UHFD world in an extensive range of so far dedicated codes for predicting molecular and crystalline properties.

It is also important and exciting for us to be able to present as much as possible experimental evidence powerfully supporting the quantitative and systematic aspects of the MGC-SDFT-UHFD method to determine the one-body energy spectra, the quasi-particle's wave functions, the magnetic property such as the mean isotropic spin-exchange coupling constants $\{J_{i,j}\}$ and the total electronic internal energies, in Section 2.3. In [13], we provided the first experimental evidence for it; in **Table 1**, the derived formulas for $J_{1,2}$ demonstrated excellent quantitative agreements (less than 1% errors) with 10 experimental results from biomimetic binuclear transition metal complexes (TM: Cu, Mn, Fe), using the Mulliken's atomic spin densities [14] and a set of the internal energies calculated by the UB3LYP/PBS/lacvp** method [15, 16]. Among many controversial problems that remained to be elucidated in photosynthesis research (see a recent review [17]), in Section 2.4 we discuss the second experimental supporting evidence provided by two broad EPR signals, named "g4.8" [18, 19] and "g12-multiline" [20], observed from the dark-stable S_1 state Mn_4CaO_5 clusters in the PSII having slightly different structures between thermophilic cyanobacteria in [18, 19] and higher-plant spinach in [20], respectively. At present, however, we have at hand only the structure of former's PSII crystal at 1.95 Å high-resolution viewed by femtosecond XFEL pulse irradiation [21] but do not have any structure of the latter PSII crystal at least at similar high-resolution. It should be also noted that the super-brilliant femtosecond XFEL-pulse irradiation may generate high-density secondary photoelectrons to deoxidize nearby Mn_4 clusters with high probability during diffraction measurements. Then, the quantitative determination of the Heisenberg spin Hamiltonian involved in *the principal exchange-correlation energy function* can play a key role in the geometry optimization by the UB3LYP/PBS(ε)/lacvp** method to make the model Mn_4CaO_x cluster being thermally distributed in some isomeric substates of any Kok-S_i state.

Furthermore, in Section 2.5, we discuss the most interesting many-body effect induced by the *secondary correlation energy term*, which represents a spin-dependent attractive correlation interaction between a couple of conductive Bloch-electrons with antiparallel spins that could be generated only near the Fermi surface in the metallic crystal. This strong correlation interaction may accelerate the phase transition from the normal state to the superconductive state by promoting Cooper-pairings of conductive Bloch-electrons near the Fermi level in the superconductor against the common knowledge [22–27].

A problematic idea underlining the KS-DFT formalism may be described in other words such that the ground state energy E of the one-particle self-consistent field Hamiltonian for N electron systems, which corresponds to the internal energy functional of the electron density determined in thermal equilibrium state, should be further

DOI: http://dx.doi.org/10.5772/intechopen.111746

Complexes	a $Cu_2^{II,II}$	c $Mn_2^{IV,IV}$	d $Mn_2^{III,IV}$	e $Mn_2^{III,IV}$	f $Mn_2^{IV,III}$	g $Mn_2^{IV,IV}$	h $Mn_2^{II,II}$	i $Fe_2^{II,II}$	j $Mn_2^{III,III}$	k $Fe_2^{III,III}$
N_A	62	118	91	89	86	88	77	77	81	81
$R_{1,2}$(Å)	2.659	2.745	2.591	2.591	3.230	2.296	3.370	3.315	3.140	3.202
bridge ligands	$(\mu-OAc)_4$	$(\mu-O)_2$	$(\mu-O)_2$ $(\mu-OAc)$	$(\mu-O)_2$ $(\mu-OAc)$	$(\mu-O)$ $(\mu-OAc)_2$	$(\mu-O)_3$	$(\mu-OH)$ $(\mu-OAc)_2$	$(\mu-OH)$ $(\mu-OAc)_2$	$(\mu-OH)$ $(\mu-OAc)_2$	$(\mu-OH)$ $(\mu-OAc)_2$
ε	10	20	20	20	40	20	40	20	20	10
$\Delta U^{(2)}_{UHFD}(cm^{-1})$	−271.1	−1366.	−1579.	−1549.	−506.6	−3928.	−248.9	−223.8	141.4	−2522.
$n^{(1)}_{1EF}$ $n^{(2)}_{1EF}$	0.975 0.980	3.064 3.033	4.040 4.031	4.087 4.111	3.140 −3.22	3.106 3.038	4.939 4.912	3.913 3.907	4.006 3.882	4.686 4.583
$n^{(1)}_{2EF}$ $n^{(2)}_{2EF}$	0.976 −0.981	3.107 −3.039	3.076 −3.189	3.034 −3.187	4.010 4.010	3.103 −3.026	4.941 −4.911	3.912 −3.907	4.006 −3.886	4.683 −4.636
$S^{(1)}_{1EF,2EF}$ $S^{(2)}_{1EF,2EF}$ $(\times 10^{-2})$	5.13 −3.76	−5.46 2.38	−3.15 5.50	−5.47 8.21	−3.74 5.69	−6.61 2.13	2.44 −3.65	4.52 −4.80	0.280 −6.06	13.94 16.46
$J^{(4-0)}_{1,2}$ (cm^{-1})	−284.9	−143.5	−126.5	−123.4	−40.1	−407.6	−10.2	−14.6	8.8	−114.9
$J^{exp}_{1,2}$ (cm^{-1})	−285	−144	−125	−125	−40	−407	−9	−14	9	−115

Table 1.
*Benchmark-test results of the 2GC-UHFD-SD averaged ES-exchange coupling constants, designated $J^{(4-0)}_{1,2}$, for 10 biomimetic binuclear Cu, Mn and Fe complexes, using the UB3LYP/PBS(ε)/lacvp** (4th XC) method combined with the inherent formulas of Eqs. (103), (113), derived in the present MGC-UHFD-SDFT (0th XC) method. These model TM_2 complexes consist of N_A atoms, are imbedded in each ε dielectric constant medium, and exhibit a variety of TM1-TM2 distances, designated $R_{1,2}$, which depend strongly on different* bridge structures *and the different* TM valences, *and also weakly on the different non-bridging ligations of paramagnetically-polarized O atoms and diamagnetically-/paramagnetically-polarized N atoms (not shown here). Here the data for* **b** *(di-μ-oxo bridged Mn^{IV}-Mn^{IV} dimer ligated by four picolinic anions) in* **Table I** *in [13] was omitted because of its optimized structure containing no solvent molecules. (see Supplemental Online Material of Ref. [13] about how to calculate the effective spin densities using the Mulliken's atomic spin densities [14].)*

minimized by "the *exact* variation principle" with respect to the electron density regarded as a variational variable to search for "the *exact* energy functional" of "the *exact* electron density", subjected to the N-representability condition [4]. This wrong variational idea appears to have been widely accepted so far, although it may have been enforced by a special situation enforced by too strong expressions involving many *exact's*: "the *exact* variational principle", and "the existence theorem of an *exact* energy functional of the *exact* electron density" as well as "N-representability condition". Especially, "the N-representability condition" seems to be too strong to consider any open quantum system, in which the total numbers (N_α, N_β) of (up, down) electrons in the system should be replaced by the mean values $(\overline{N}_\alpha, \overline{N}_\beta)$ of a pair of the expected values of their operators, $(\widehat{N}_\alpha, \widehat{N}_\alpha)$, respectively, in the context of applying the variational principle to the minimum grand potential including them, as will be shown in this chapter. So far, neither theoretical proof nor evidence for the K-S formalism could not be provided unless the *exact* correlation energy function is discovered.

In a GC ensemble, one may consider a much larger M-electron system ($M \gg N$) of atoms, molecules, and solids, which will be maximally realized with a finite probability in contact with a grand canonical heat/particle reservoir containing a much larger number of electrons at temperature θ/k_B (k_B is the Boltzmann constant). All the stational states of the M-electron system, which involve the ground and all kinds of excited states, may be assumed to be describable in terms of the time-independent non-relativistic Schrödinger wave equation in 3D-space:

$$H\Psi = E\Psi(\boldsymbol{x}_1, \boldsymbol{x}_2, \cdots\cdots, \ \boldsymbol{x}_M); M \gg N, \tag{1}$$

where H is the Hamiltonian operator given by

$$H = T + V_{ne} + V_{ee}, \tag{2}$$

$$= \sum_{i=1}^{M}\left(-\frac{1}{2}\nabla_i^2\right) + \sum_{i=1}^{M} \upsilon(\boldsymbol{r}_i, \varepsilon) + \sum_{i<j}^{M} \frac{1}{r_{i,j}}. \tag{3}$$

Here, $\boldsymbol{x}_i \equiv (\boldsymbol{r}_i, \boldsymbol{s}_i)$ represents the (orbital, spin) coordinates of the ith electron, T is the electron kinetic energy operator; V_{ne} is the electrostatic interaction operator of electrons with all nuclei and the surrounding medium of the dielectric constant ε if a convenient "the linear Poisson-Boltzmann equation Solver (PBS)" model [8] is augmented; V_{ee} is the electron Coulomb interaction operator; and $r_{i,j} \equiv |\boldsymbol{r}_i\text{-}\boldsymbol{r}_j|$. Since it is impossible to exactly solve Eq. (1) except for the case of a hydrogen atom, we have developed the ultimate MGC-SDFT formalism [7], which has been constructed by developing five new methodological concepts in Subsections 2.1 through 2.5 along the basic principles of quantum thermodynamics with the theory of open quantum systems, but not of closed quantum system as adopted in the Kohn-Sham formalism.

2. Multiple grand canonical spin density functional theory

2.1 Definition of a grand canonical ensemble: One-particle and two-particle reduced density matrices

The principally most general choice would be made for an *extended* antisymmetric Slater determinant wave function as the trial many-electron wave function $\Psi(\boldsymbol{x}_1,$

$\boldsymbol{x}_2, \dots, \boldsymbol{x}_M)$ in the Schrödinger Eq. (1), consisting of a complete large enough number, (M_α, M_β), of mutually-independent *natural molecular spin-orbitals* (NMO) wave functions, Eq. (6), which may be most-appropriately expandable in terms of a Linear Combination of gaussian-type or Slater-type Atomic Orbitals (LCAO), $\mathbf{G}_A \equiv [g_l^a(\boldsymbol{r})]$ (a, the atomic order number from 1 to N_A; l, a set of plural AO quantum numbers). The maximum size of (M_α, M_β)-dimensional Hilbert space of the NMO set $\mathbf{M}'$ of Eq. (4) must be as large enough as possible to satisfy the near-completeness condition of Eq. (5) and the orthonormality relations of Eqs. (6) and (7) as far as the highest NMO energy levels may not exceed a dissociation limit given by a work function, W.

$$\mathbf{M}' = {}^\alpha\mathbf{M}' + {}^\beta\mathbf{M}' \equiv \left\{\Psi'_\tau(\mathbf{x}), \epsilon'_\tau; \tau = 1_\alpha, \dots, M_\alpha, 1_\beta, \dots, M_\beta\right\}, \tag{4}$$

$$\sum_{m\sigma}^{{}^\sigma\mathbf{M}'} |\Psi'_{m\sigma}\rangle\langle\Psi'_{m\sigma}| \cong \mathbf{1} \text{ for } M_\sigma >> N_\sigma; \sigma = \alpha, \beta. \tag{5}$$

$$\Psi'_{m\sigma}(\boldsymbol{x}) \equiv \Phi'_{m\sigma}(\boldsymbol{r})\xi'_{m\sigma}(\boldsymbol{s}), \tag{6}$$

$$\langle\Psi'_{m\sigma}|\Psi'_{m\sigma''}\rangle = \langle\Phi'_{m\sigma}|\Phi'_{m\sigma''}\rangle\langle\xi'_{m\sigma}|\xi'_{m\sigma''}\rangle = \delta_{m\sigma,m\sigma''}\delta_{\sigma,\sigma''}, \tag{7}$$

$$\Phi_{m\sigma}(\boldsymbol{r}) = \sum_{a,l}^{\mathbf{G_A}} C_{a,l}^{m\sigma} g_l^a(\boldsymbol{r}) = \sum_{a=1}^{N_A}\sum_{l} C_{a,l}^{m\sigma} g_l^a(\boldsymbol{r}). \tag{8}$$

In Eq. (8), the AO basis functions, which may include polarization and diffuse functions, are assumed to be orthonormalized in each atom such as $\langle g_l^a | g_{l'}^a\rangle = \delta_{l,l'}$ but slightly overlap between the valence-electron orbitals of neighboring atoms, except for most AOs in the core levels. Hereafter, the single dashed quantities in Eqs. (4)–(7) will be used for the quantities in a thermal nonequilibrium state. The completeness Eq. (5) and the orthonormality Eqs. (6) and (7) are assumed to seamlessly hold even in such non-equilibrium states.

It is important to remind that the thermodynamic equilibrium state can be achieved in terms of the Rayleigh-Ritz variational principle applied to the one-particle reduced density matrix given by

$$\widehat{\Gamma}[p', \Psi'] \equiv \sum_{N'}^{\infty}\sum_{\tau}^{\mathbf{M}'} P'_{N',\tau}|\Psi'_{N',\tau}\rangle\langle\Psi'_{N',\tau}|, \tag{9}$$

$$= \sum_{\tau}^{\mathbf{M}'}\sum_{n_\tau=0}^{1} p'_{\tau,n_\tau}|\Psi'_{\tau,n_\tau}\rangle\langle\Psi'_{\tau,n_\tau}|, \tag{10}$$

with respect to a set of the distribution probabilities, designated $\{P'_{N',\tau}; N' = 0, \dots, \infty; \tau = 1, \dots, M\}$ for fermions and bosons in Eq. (9) or $\{p'_{\tau,n_\tau}; \tau = 1, \dots, \mathrm{M}; n_\tau = 0, 1\}$ for (NMO-transformed) fermions in Eq. (10). This nonequilibrium state will relax to the maximum entropy state keeping the normalization condition of Eq. (11) and the binary chemical potential (μ_α/μ_β) equilibrium conditions with the heat/particle reservoir leading to Eqs. (12) and (13):

$$\mathrm{Tr}\left[\widehat{\Gamma}\right] = \sum_{\tau}^{\mathbf{M}'}\sum_{n_\tau=0}^{1} p'_{\tau,n_\tau} = 1, \tag{11}$$

$$\widehat{N}_\sigma = \sum_{\tau(\sigma)}^{{}^\sigma\mathbf{M}'} \widehat{a}^\dagger_{\tau(\sigma)}\widehat{a}_{\tau(\sigma)}; \sigma = \alpha, \beta \tag{12}$$

$$\mathrm{Tr}\left[\widehat{N}_\sigma \widehat{\Gamma}\right] = \sum_{\tau(\sigma)}^{{}^\sigma\mathbf{M}'} \sum_{n_{\tau(\sigma)}=0}^{1} n_{\tau(\sigma)} p'_{\tau(\sigma),n_{\tau(\sigma)}} \equiv \overline{N}_\sigma[p', \Psi'], \tag{13}$$

where $\widehat{a}_\tau$ and $\widehat{a}^\dagger_\tau$ are the annihilation and creation operators of an electron in the τth NMO-eigenstate, respectively. Subsequently, using the GC entropy

$$\Sigma[p', \Psi'] = -k_B \mathrm{Tr}\left(\widehat{\Gamma}\ \ln \widehat{\Gamma}\right) = -k_B \sum_{\tau}^{\mathbf{M}'} \sum_{n_\tau=0}^{1} p'_{\tau,n_\tau} \ln p'_{\tau,n_\tau}, \tag{14}$$

and the second quantization expression of the Hamiltonian operator, that is

$$\widehat{H} = \widehat{H}_0 + \widehat{H}_1 = \sum_{\tau}^{\mathbf{M}'} \epsilon'_\tau \widehat{a}^\dagger_\tau \widehat{a}_\tau + \widehat{H}_1, \tag{15}$$

which in general consists of the principal part $\widehat{H}_0$ and the secondary part $\widehat{H}_1$, responsible for the GC ensemble of mutually independent NMO fermions and a perturbational interaction between them, respectively, we obtain the grand potential in thermal nonequilibrium state:

$$\Omega[p', \Psi'] \equiv \mathrm{Tr}\left[\widehat{\Gamma}\left(\theta \ln \widehat{\Gamma} + \widehat{H}_0 - \mu_\alpha \widehat{N}_\alpha - \mu_\beta \widehat{N}_\beta\right)\right], \tag{16}$$

$$= \sum_{\tau}^{\mathbf{M}'} \sum_{n_\tau=0}^{1} p'_{\tau,n_\tau} \left[\theta \ln p'_{\tau,n_\tau} + \langle \Psi'_\tau, n_\tau | \left(\epsilon'_\tau - \mu_{\sigma(\tau)}\right) n_\tau | \Psi'_\tau, n_\tau \rangle\right]. \tag{17}$$

In Eqs. (16) and (17) should be noted that $\widehat{H}$ *is replaced by the principal part* $\widehat{H}_0 = \sum_{\tau}^{\mathbf{M}'} \epsilon'_\tau \widehat{a}^\dagger_\tau \widehat{a}_\tau$. The variational equations subject to the normalization condition Eq. (11) are given by

$$\frac{\partial}{\partial p'_{\tau,n_\tau}} \left(\Omega[\{p'_{\tau,n_\tau}, \Psi'_\tau\}] + \lambda \left[\sum_{\tau}^{\mathbf{M}} \sum_{n_\tau=0}^{1} p'_{\tau,n_\tau} - 1\right]\right) = 0, \tag{18}$$

with λ being a Lagrange's multiplier, satisfying

$$e^{-1-\lambda/\theta} = \sum_{\tau}^{\mathbf{M}'} \sum_{n_\tau=0}^{1} \langle \Psi'_\tau, n_\tau | \epsilon'_\tau - \mu_\alpha \overline{N}_\alpha - \mu_\beta \overline{N}_\beta | \Psi'_\tau, n_\tau \rangle. \tag{19}$$

Thus, we obtain an intermediate solution with fixed $\{\Psi'_\tau\}$:

$$p'_{\tau,n_\tau} \overset{yields}{\rightarrow} p^0_{\tau,n_\tau}(\Psi') = \frac{\exp\left[\langle \Psi'_\tau, n_\tau | \left(\mu_{\sigma(\tau)} - \epsilon'_\tau\right) n_\tau | \Psi'_\tau, n_\tau \rangle / \theta\right]}{\sum_{n_\tau=0}^{1} \exp\left[\langle \Psi'_\tau, n_\tau | \left(\mu_{\sigma(\tau)} - \epsilon'_\tau\right) n_\tau | \Psi'_\tau, n_\tau \rangle / \theta\right]} \tag{20}$$

$$\widehat{\Gamma}^0(\Psi') = \sum_{\tau}^{\mathbf{M}'} \sum_{n_\tau=0}^{1} p^0_{\tau,n_\tau}(\Psi') \lfloor \Psi'_{\tau,n_\tau} \rangle \langle \Psi'_{\tau,n_\tau} |. \tag{21}$$

Then, the GC potential decreases by the non-negative quantity (the equality appears when $p' = p^0$), namely

$$\begin{aligned} &\Omega[\{p'_{\tau,n_\tau}, \Psi'_\tau\}] - \Omega\left[\left\{p^0_{\tau,n_\tau}, \Psi'_\tau\right\}\right] \\ &= \sum_{\tau}^{\mathbf{M}} \sum_{n_\tau=0}^{1} p'_{\tau,n_\tau}\left(\theta \ln p'_{\tau,n_\tau} + \langle \Psi'_\tau, n_\tau | \left(\epsilon'_\tau - \mu_{\sigma(\tau)}\right) n_\tau | \Psi'_\tau, n_\tau \rangle\right) \\ &+\theta \ln \sum_{\tau}^{\mathbf{M}} \sum_{n_\tau=0}^{1} \exp\left[\langle \Psi'_\tau, n_\tau | \left(\mu_{\sigma(\tau)} - \epsilon'_\tau\right) n_\tau | \Psi'_\tau, n_\tau \rangle / \theta\right] \\ &= \theta \sum_{\tau}^{\mathbf{M}} \sum_{n_\tau=0}^{1} p'_{\tau,n_\tau}\left[\ln p'_{\tau,n_\tau} - \ln p^0_{\tau,n_\tau}(\Psi')\right] \geq 0, \end{aligned} \tag{22}$$

which is notably induced only by the increase of the GC entropy $\Sigma\ [p',\Psi']\ \text{-}\Sigma\ [p^0,\Psi']$. This initial-guess state of the GC ensemble $\mathbf{M}' = \{\Psi'_\tau, \epsilon'_\tau\}$ will relax to converge toward the self-consistent, orthonormal and complete eigenvectors/eigenvalues set **M** of Eqs. (23)–(27) by iteration technique, which will be described in Section 2.2.

$$\mathbf{M} = {}^{\alpha}\mathbf{M} + {}^{\beta}\mathbf{M} \equiv \left\{\Psi_{m\sigma}(\mathbf{x}), \epsilon_{m\sigma}, f_{m\sigma}(\mu_\sigma); m\sigma \equiv \tau = 1_\alpha, \dots, \mathrm{M}_\alpha, 1_\beta, \dots, \mathrm{M}_\beta\right\}, \tag{23}$$

$$\sum_{m\sigma}^{{}^{\sigma}\mathbf{M}} |\Psi_{m\sigma}\rangle\langle\Psi_{m\sigma}| \cong \mathbf{1} \text{ for } \mathrm{M}_\sigma > > \mathrm{N}_\sigma; \sigma = \alpha, \beta. \tag{24}$$

$$\Psi_{m\sigma}(\boldsymbol{x}) \equiv \Phi_{m\sigma}(\boldsymbol{r})\xi_{m\sigma}(\boldsymbol{s}), \tag{25}$$

$$\langle \Psi_{m\sigma} | \Psi_{m\sigma'} \rangle = \langle \Phi_{m\sigma} | \Phi_{m\sigma'} \rangle \langle \xi_{m\sigma} | \xi_{m\sigma'} \rangle = \delta_{m\sigma,m\sigma''} \delta_{\sigma,\sigma''}, \tag{26}$$

$$f_{m\sigma}(\mu_\sigma) = p_{m\sigma,1} = \{\exp[(\epsilon_{m\sigma} - \mu_\sigma)/\theta] + 1\}^{-1} \equiv f(\epsilon_{m\sigma} - \mu_\sigma), \tag{27}$$

where it should be noted that only the populated distribution probability $p^0{}_{m\sigma,1}$ of Eq. (20) converges to the Fermi-Dirac distribution probability $f_{m\sigma}(\mu_\sigma)$ of Eq. (27).

Next, we need to introduce the first-order (for one-particle interactions) and the second order (for two-particle interactions) reduced electron density matrixes, $\Gamma(\boldsymbol{x}, \boldsymbol{x}')$ and $\Gamma_2(\boldsymbol{x}_1\boldsymbol{x}_2, \boldsymbol{x}'_1\boldsymbol{x}'_2)$, respectively, for the GC ensemble **M** in thermal equilibrium state, as given by

$$\Gamma(\boldsymbol{x}, \boldsymbol{x}') = \sum_{\tau}^{\mathbf{M}} f_\tau |\Psi_\tau(\boldsymbol{x})\rangle\langle\Psi_\tau(\boldsymbol{x}')|, \tag{28}$$

$$\Gamma_2\left(\boldsymbol{x}_1\boldsymbol{x}_2, \boldsymbol{x}'_1\boldsymbol{x}'_2\right) = \frac{1}{2}[\Gamma(\boldsymbol{x}_1, \boldsymbol{x}'_1)\Gamma(\boldsymbol{x}_2, \boldsymbol{x}'_2) - \Gamma(\boldsymbol{x}_1, \boldsymbol{x}'_2)\Gamma(\boldsymbol{x}_2, \boldsymbol{x}'_1)], \tag{29}$$

which will be used in Section 2.2.

2.2 The self-consistent field method in the UHFD approximation

Here, we derive the self-consistent field (SCF) method to generate such a realistic GC ensemble, **M**, as given by Eqs. (23)–(27), in which all NMO levels will be partially occupied with the Fermi-Dirac distribution probability $f(\epsilon_{m\sigma}-\mu_\sigma)$ constrained by the chemical potentials μ_σ defined by either Eq. (27) or the Gibbs free energy per a σ-spin electron for σ = (α, β), as given by Eq. (30), using the mean number of σ-spin NMO-fermions for σ = (α, β) in Eq. (31):

$$\mathrm{G} = \mu_\alpha \overline{N}_\alpha + \mu_\beta \overline{N}_\beta, \tag{30}$$

$$\overline{N}_\sigma = \sum_{m\sigma}^{^\sigma\mathbf{M}} f(\epsilon_{m\sigma} - \mu_\sigma). \tag{31}$$

At first, we define various spinless electron density matrixes for later usage:

$$\rho_\sigma(\boldsymbol{r}, \boldsymbol{r}') = \mathrm{Tr}_s[\Gamma_\sigma(\boldsymbol{x}, \boldsymbol{x}')]_{s=s'} = \sum_{m\sigma}^{^\sigma\mathbf{M}} f_{m\sigma}(\mu_\sigma)\Phi_{m\sigma}(\boldsymbol{r})\Phi^*_{m\sigma}(\boldsymbol{r}'); \sigma = \alpha, \beta, \tag{32}$$

$$\rho(\boldsymbol{r}, \boldsymbol{r}') = \rho_\alpha(\boldsymbol{r}, \boldsymbol{r}') + \rho_\beta(\boldsymbol{r}, \boldsymbol{r}'), \tag{33}$$

$$\rho_\sigma(\boldsymbol{r}) \equiv \rho_\sigma(\boldsymbol{r}, \boldsymbol{r}), \rho(\boldsymbol{r}) = \rho_\alpha(\boldsymbol{r}) + \rho_\beta(\boldsymbol{r}), \tag{34}$$

where Tr_s represents the trace on the spin coordinate s.

Next, we will prove that the internal energy $U[\Gamma_\alpha, \Gamma_\beta]$ as a function of the reduced density matrix Γ = (Γ_α, Γ_β) can be decoupled into two parts, as seen in Eq. (36): (1) the principal part $U^0_{UHFD}[\Gamma_\alpha, \Gamma_\beta]$ of Eq. (37) including the principal exchange-correlation energy functional -K_{XC} [ρ_α, ρ_β] defined by Eq. (40) and (2) a secondary part containing only a spin-dependent correlation energy functional $\Delta E^{corr}_{UHFD}[\Gamma_\alpha, \Gamma_\beta]$ defined by Eq. (43). This ultimate decoupling scheme neglecting the secondary correlation term ΔE^{corr}_{UHFD} [Γ_α, Γ_β] of Eq. (43) will be tentatively called "the *Unrestricted Hartree-Fock-Dirac (UHFD) approximation*". because the Dirac's spin permutation operator ($\boldsymbol{\sigma}_i$ is called spinor)

$$P^\sigma_{12} = \frac{1}{2}[1 + (\boldsymbol{\sigma}_1, \boldsymbol{\sigma}_2)] = \frac{1}{2}[1 + 4(\boldsymbol{s}_1, \boldsymbol{s}_2)], \tag{35}$$

including an inner product of two of Pauli's spin operators $(\sigma_1, \sigma_2) = (2s_1, 2s_2)$, has played a decisive role in our discovery of this new decoupling scheme. This is indeed a revolutionary discovery a long way beyond the early Unrestricted Hartree (UH), Unrestricted Hartree-Fock (UHF) and Unrestricted Hartree-Fock-Slater (UHFS) approximations. This new UHFD decoupling scheme leads to a group of fundamental equations:

$$U[\Gamma_\alpha, \Gamma_\beta] = U^0_{UHFD}[\Gamma_\alpha, \Gamma_\beta] + \Delta E^{corr}_{UHFD}[\Gamma_\alpha, \Gamma_\beta], \tag{36}$$

$$U^0_{UHFD}[\Gamma_\alpha, \Gamma_\beta] \equiv \mathrm{T}[\rho] + V_{ne}[\rho] + J[\rho] - \frac{1}{2}K_{XC}[\rho_\alpha, \rho_\beta] + H_{ES}[\Gamma_\alpha, \Gamma_\beta], \tag{37}$$

$$V_{ne}[\rho] = \int d\boldsymbol{r} v(\boldsymbol{r}, \varepsilon)\rho(\boldsymbol{r}), \tag{38}$$

$$J[\rho] = \frac{1}{2}\iint \frac{d\mathbf{r}_1 d\mathbf{r}_2}{r_{12}} \rho(\mathbf{r}_1)\rho(\mathbf{r}_2), \tag{39}$$

$$-K_{XC}[\rho_\alpha, \rho_\beta] = -\frac{1}{2}\iint \frac{d\mathbf{r}_1 d\mathbf{r}_2}{r_{12}} \rho(\mathbf{r}_1, \mathbf{r}_2)\rho(\mathbf{r}_2, \mathbf{r}_1), \tag{40}$$

$$H_{ES}[\Gamma_\alpha, \Gamma_\beta] = -2\iint \frac{d\mathbf{r}_1 d\mathbf{r}_2}{r_{12}} Q(\mathbf{r}_1, \mathbf{r}_2)Q(\mathbf{r}_2, \mathbf{r}_1)(\mathbf{s}_1, \mathbf{s}_2), \tag{41}$$

$$Q(\mathbf{r}_1, \mathbf{r}_2) = \rho_\alpha(\mathbf{r}_1, \mathbf{r}_2) - \rho_\beta(\mathbf{r}_1, \mathbf{r}_2), \tag{42}$$

$$\Delta E^{corr}_{UHFD}[\Gamma_\alpha, \Gamma_\beta] = -2\iint \frac{d\mathbf{r}_1 d\mathbf{r}_2}{r_{12}} [\rho_\alpha(\mathbf{r}_1, \mathbf{r}_2)\rho_\beta(\mathbf{r}_2, \mathbf{r}_1) + \rho_\beta(\mathbf{r}_1, \mathbf{r}_2)\rho_\alpha(\mathbf{r}_2, \mathbf{r}_1)](\mathbf{s}_1, \mathbf{s}_2), \tag{43}$$

where $T[\rho]$ represents the expected value of the electron kinetic energy operator T, although this notation does not mean any explicit functional form of ρ, the other explicit energy functionals of ρ have usual meanings, and $H_{ES}[\Gamma_\alpha, \Gamma_\beta]$ does the spin density coupling energy functional between two NMO-fermions, which is expected to contain H_{ex}. (CORREGENDUM: Please add a miss-dropped factor, 2, in Eq. (3.18) in [13], just like above Eq. (41)).

[Proof of Eqs. (36)–(42)] Substituting Eq. (28) into Eq. (29), we get the NMO expansion formula of the exchange-product matrix:

$$\Gamma(\mathbf{x}_1, \mathbf{x}'_2)\Gamma(\mathbf{x}_2, \mathbf{x}'_1) = \sum_{\sigma, m\sigma}^{\mathbf{M}} f_{m\sigma}(\mu_\sigma)\Phi_{m\sigma}(\mathbf{r}_1)\Phi_{m\sigma}(\mathbf{r}'_2)^* \\ \times \sum_{\sigma', m\sigma'}^{\mathbf{M}} f_{m\sigma'}(\mu_{\sigma'})\Phi_{m\sigma'}(\mathbf{r}_2)\Phi_{m\sigma}(\mathbf{r}'_1)^* |\xi_{m\sigma}(\mathbf{s}_2)\xi_{m\sigma'}(\mathbf{s}_1)\rangle\langle\xi_{m\sigma}(\mathbf{s}'_1)\xi_{m\sigma'}(\mathbf{s}'_2)|. \tag{44}$$

Here, to restore the spin-pair wave function to the normal-order form, Dirac's spin operator of Eq. (35) needs to be operated to the two-spin function:

$$\xi_{m\sigma}(\mathbf{s}_2)\xi_{m\sigma'}(\mathbf{s}_1) = P^{\sigma}_{12}\xi_{m\sigma}(\mathbf{s}_1)\xi_{m\sigma'}(\mathbf{s}_2). \tag{45}$$

Then, using Eqs. (32), (33), (35), and (45), we can transform Eq. (44) into two different formulas:

$$\Gamma(\mathbf{x}_1, \mathbf{x}'_2)\Gamma(\mathbf{x}_2, \mathbf{x}'_1) \\ = \rho(\mathbf{r}_1, \mathbf{r}'_2)\rho(\mathbf{r}_2, \mathbf{r}'_1)\left[\frac{1}{2}(1 + \sigma_{1z}\sigma_{2z}) + \frac{1}{4}(\sigma_{1+}\sigma_{2-} + \sigma_{1-}\sigma_{2+})\right], \tag{46}$$

$$= \frac{1}{2}[\rho_\alpha(\mathbf{r}_1, \mathbf{r}'_2)\rho_\alpha(\mathbf{r}_2, \mathbf{r}'_1) + \rho_\beta(\mathbf{r}_1, \mathbf{r}'_2)\rho_\beta(\mathbf{r}_2, \mathbf{r}'_1)] \\ + \frac{1}{2}[\rho_\alpha(\mathbf{r}_1, \mathbf{r}'_2)\rho_\beta(\mathbf{r}_2, \mathbf{r}'_1) + \rho_\beta(\mathbf{r}_1, \mathbf{r}'_2)\rho_\alpha(\mathbf{r}_2, \mathbf{r}'_1)] \\ + \frac{1}{2}Q(\mathbf{r}_1, \mathbf{r}'_2)Q(\mathbf{r}_2, \mathbf{r}'_1)(\boldsymbol{\sigma}_1, \boldsymbol{\sigma}_2) \\ + \frac{1}{2}[\rho_\alpha(\mathbf{r}_1, \mathbf{r}'_2)\rho_\beta(\mathbf{r}_2, \mathbf{r}'_1) + \rho_\alpha(\mathbf{r}_1, \mathbf{r}'_2)\rho_\beta(\mathbf{r}_2, \mathbf{r}'_1)](\boldsymbol{\sigma}_1, \boldsymbol{\sigma}_2), \tag{47}$$

with the use of the spin density matrix Q of Eq. (42) and the off-diagonal spinors $\sigma_{j\pm}$ of Eq. (48):

$$\sigma_{j\pm} = \sigma_{jx} \pm i\sigma_{jy}, j = 1, 2. \tag{48}$$

Apparently, there exist two decoupling schemes: (1) In Eq. (46) is decoupled a pair of off-diagonal spinor terms, leading to the UHF approximation, and (2) In Eq. (47) decoupled only the last term, leading to the UHFD approximation. In the UHFD approximation, we obtain the functional formula for the internal energy $U[\Gamma_\alpha, \Gamma_\beta]$ in 3D-spin space decomposed into the principal part $U^0_{UHFD}[\Gamma_\alpha, \Gamma_\beta]$ and the secondary part $\Delta E_{UHFD}{}^{corr}$ in Eq. (36). (QED)

Since all the DFT calculations can be made in the binary-spin Hilbert space, we must take the trace of $H_{ES}\,[\Gamma_\alpha, \Gamma_\beta]$ in Eq. (37) on the 3D-spin coordinates to obtain its functional of $\rho(\rho_\alpha, \rho_\beta)$, which is equal to a half of the principal exchange-correlation energy functional, that is

$$H_{ES}[\rho_\alpha, \rho_\beta] \equiv \mathrm{Tr}_S(H_{ES}[\Gamma_\alpha, \Gamma_\beta]) = -\frac{1}{2}K_{XC}[\rho_\alpha, \rho_\beta]. \tag{49}$$

Using Eq. (49), we also obtain the principal internal energy functional of ρ:

$$U^0_{UHFD}[\rho_\alpha, \rho_\beta] \equiv \mathrm{Tr}_S(U^0_{UHFD}[\Gamma_\alpha, \Gamma_\beta]) = \mathrm{T}[\rho] + V_{ne}[\rho] + J[\rho] - K_{XC}[\rho_\alpha, \rho_\beta], \tag{50}$$

which should be equated to the GC ensemble average of the principal part of the Hamiltonian operator in the second quantization representation in the thermal equilibrium state (see Eq. (15)), that is

$$\widehat{H}^0_{UHFD} = \sum_\tau^{\mathbf{M}} \epsilon_\tau \widehat{a}^\dagger_\tau \widehat{a}_\tau, \tag{51}$$

leading to

$$U^0_{UHFD}[\rho_\alpha, \rho_\beta] \equiv \mathrm{Tr}_{\mathbf{M}}\left[\Gamma\widehat{H}_0\right] = \int d\boldsymbol{r} \sum_\tau^{\mathbf{M}} \epsilon_\tau f_\tau\left(\epsilon_\tau - \mu_{\sigma(\tau)}\right)\Phi^*_\tau(\boldsymbol{r})\Phi_\tau(\boldsymbol{r}). \tag{52}$$

Similarly, U^0_{UHFD} can be expanded as the GC ensemble average of a self-consistent effective Hamiltonian as given by

$$\begin{aligned} U^0_{UHFD}[\rho_\alpha, \rho_\beta] = \int d\boldsymbol{r} \Bigg\{ \sum_\tau^{\mathbf{M}} f_\tau\left(\epsilon_\tau - \mu_{\sigma(\tau)}\right)\Phi^*_\tau(\boldsymbol{r}) \\ \times\left[-\frac{1}{2}\nabla^2\Phi_\tau(\boldsymbol{r}) + \int \upsilon_{NMO}(\boldsymbol{r}, \boldsymbol{r}')\Phi_\tau(\boldsymbol{r}')d\boldsymbol{r}'\right]\Bigg\} \end{aligned}, \tag{53}$$

with the use of the local and non-local NMO-based effective potential defined by

$$\upsilon_{NMO}(\boldsymbol{r}, \boldsymbol{r}') = \left[\upsilon(\boldsymbol{r}, \varepsilon) + \frac{1}{2}\int d\boldsymbol{r}'' \frac{\rho(\boldsymbol{r}'')}{|\boldsymbol{r} - \boldsymbol{r}''|}\right]\delta(\boldsymbol{r} - \boldsymbol{r}') - \frac{\rho(\boldsymbol{r}, \boldsymbol{r}')}{2|\boldsymbol{r} - \boldsymbol{r}'|}, \tag{54}$$

$$\rho(\boldsymbol{r}) = \sum_\tau^{\mathbf{M}} f_\tau\left(\epsilon_\tau - \mu_{\sigma(\tau)}\right)|\Phi_\tau(\boldsymbol{r})|^2, \tag{55}$$

$$\rho(\boldsymbol{r}, \boldsymbol{r}') = \sum_\tau^{\mathbf{M}} f_\tau\left(\epsilon_\tau - \mu_{\sigma(\tau)}\right)\Phi^*_\tau(\boldsymbol{r}')\Phi_\tau(\boldsymbol{r}). \tag{56}$$

From equivalent Eqs. (52) and (53) leading to Eq. (57), we obtain a series of central Schrödinger equations, (58) by putting $[\dots]_\tau = 0$:

$$U^0_{UHFD}[\rho_\alpha, \rho_\beta] - \mathrm{Tr}_{\mathbf{M}}\left[\Gamma \widehat{H}_0\right] = \int dr \left\{ \sum_\tau^{\mathbf{M}} f_\tau\left(\epsilon_\tau - \mu_{\sigma(\tau)}\right) \Phi^*_\tau(\boldsymbol{r}) \right.$$
$$\left. \left[-\frac{1}{2}\nabla^2 \Phi_\tau(\boldsymbol{r}) + \int \upsilon_{NMO}(\boldsymbol{r}, \boldsymbol{r}')\Phi_\tau(\boldsymbol{r}')d\boldsymbol{r}' - \epsilon_\tau \Phi_\tau(\boldsymbol{r}) \right] \right\} = 0 \quad (57)$$

$$-\frac{1}{2}\nabla^2 \Phi_\tau(\boldsymbol{r}) + \int d\boldsymbol{r}' \upsilon_{NMO}(\boldsymbol{r}, \boldsymbol{r}')\Phi_\tau(\boldsymbol{r}') = \epsilon_\tau \Phi_\tau(\boldsymbol{r}), \quad (58)$$
$$\text{and h.c. for all } \tau \in \mathbf{M}.$$

This central (not variational!) solution will be stocked as the presumed GC ensemble $\mathbf{M}$, in which the eigenvalues $\{\epsilon_\tau\}$ are usually assumed to be rearranged from the minimum $\epsilon_{1\sigma}$ to the maximum $\epsilon_{M\sigma}$ in the order of increasing energies, first for α-spin NMOs and second for β-spin NMOs, as

$$\tau = 1_\alpha, 2_\alpha, \cdots, M_\alpha, 1_\beta, 2_\beta, \cdots, M_\beta = 1, 2, \cdots, M \left(M = M_\alpha + M_\beta\right). \quad (59)$$

For simplicity, we assume that there exists no degeneracy in energy levels in the unrestricted large system without any structural symmetry.

On the other hand, the secondary correlation energy functional, $\Delta E^{corr}_{UHFD}[\Gamma_\alpha, \Gamma_\beta]$, defined by Eq. (43), represents the sole perturbation term in 3D-spin space. Taking the trace of it on the 3D-spin coordinates, we obtain the second quantization expression of it as follows

$$\widehat{H}^1_{UHFD} = \mathrm{Tr}_s\left(\Delta E^{corr}_{UHFD}[\Gamma_\alpha, \Gamma_\beta]\right)$$
$$= -\frac{1}{2}\iint \frac{d\boldsymbol{r}_1 d\boldsymbol{r}_2}{r_{12}} \mathrm{Tr}_s\left\{ \left[\rho_\alpha(\boldsymbol{r}_1, \boldsymbol{r}_2)\rho_\beta(\boldsymbol{r}_2, \boldsymbol{r}_1) + \rho_\beta(\boldsymbol{r}_1, \boldsymbol{r}_2)\rho_\alpha(\boldsymbol{r}_2, \boldsymbol{r}_1)\right](\boldsymbol{\sigma}_1, \boldsymbol{\sigma}_2) \right\}$$
$$= -\sum_{m\alpha}^{{}^\alpha\mathbf{M}} \sum_{m\beta}^{{}^\beta\mathbf{M}} f(\epsilon_{m\alpha} - \mu_\alpha) f(\epsilon_{m\beta} - \mu_\beta) \iint \frac{d\boldsymbol{r}_1 d\boldsymbol{r}_2}{r_{12}}$$
$$\times \left[\Phi^*_{m\alpha}(\boldsymbol{r}_2)\Phi_{m\alpha}(\boldsymbol{r}_1)\Phi^*_{m\beta}(\boldsymbol{r}_1)\Phi_{m\beta}(\boldsymbol{r}_2)\right]\left(\sigma_{+,m\alpha,}\sigma_{-,m\beta} + \sigma_{-,m\alpha}\sigma_{+,m\beta}\right) \quad (60)$$
$$= -\sum_{m\alpha}^{{}^\alpha\mathbf{M}} \sum_{m\beta}^{{}^\beta\mathbf{M}} \left(\sigma_{+,m\alpha,}\sigma_{-,m\beta} + \sigma_{-,m\alpha}\sigma_{+,m\beta}\right) \iint \frac{d\boldsymbol{r}_1 d\boldsymbol{r}_2}{r_{12}}$$
$$\times \left[\Phi^*_{m\alpha}(\boldsymbol{r}_2)\Phi^*_{m\beta}(\boldsymbol{r}_1)\Phi_{m\beta}(\boldsymbol{r}_2)\Phi_{m\alpha}(\boldsymbol{r}_1)\right] \widehat{a}^\dagger_{m\beta}\widehat{a}_{m\beta}\widehat{a}^\dagger_{m\sigma}\widehat{a}_{m\alpha}.$$

The first-order perturbation term vanishes owing to the nondiagonal spinors. However, the second-order perturbation correction can always induce a finite attractive force between any pair of NMO-fermions with antiparallel spins. The most interesting example would be a positive enhancement effect on the Cooper-pair superconductivity due to an additional attractive force between two conductive Bloch-electrons with antiparallel spins near Fermi level, as will be discussed in Section 2.4.

2.3 MGC-SDFT-UHFD method for polynuclear transition metal complexes

We next consider a variety of paramagnetic systems including plural n (≥ 2) spins, designated $\{S_i, i = 1, \ldots, n\}$, which arise from transition metal (TM) cations, C/N/O-radicals, -C=C- bond radicals, and so on. These spins are quantum-mechanically interacting with each other via the exchange coupling constants ($J_{i,j}$) in the Heisenberg spin Hamiltonian defined by

$$H_{ex} = -2\sum_{i=1}^{n-1}\sum_{j=i+1}^{n} J_{i,j}\left(S_i, S_j\right). \tag{61}$$

However, this H_{ex} model takes account only the pure spin operators $\{S_i\}$ but does not contain any kind of polarized spins of ligand atoms, designated $\{s_{iL}\}$, "Why?" The most fundamentally important question is, "What is the origin of H_{ex}?" These questions have been recently solved by Kusunoki [13], as reviewed in this subsection. Let us investigate what kinds of spin-dependent physical processes are involved in the spin-density coupling energy functional $H_{ES}\,[\Gamma_\alpha,\Gamma_\beta]$ of Eqs. (41) and (42).

Since in the binary spin space appear 2^{n-1} ($n \geq 2$) mutually-independent up/down-*i*ES arrangements (ESA), which represent one ferromagnetic and the other anti-ferromagnetic states, we must prepare a set of multiple grand canonical (MGC) ensembles, as given by

$$\mathbf{M} = \sum_{k=1}^{2^{n-1}} \mathbf{M}^{(k)}, \mathbf{M}^{(k)} = {}^{\alpha}\mathbf{M}^{(k)} + {}^{\beta}\mathbf{M}^{(k)}; \tag{62}$$

$${}^{\sigma}\mathbf{M}^{(k)} \equiv \left\{\Psi_{m\sigma}^{(k)}(\mathbf{x}), \epsilon_{m\sigma}^{(k)}, f_{m\sigma}^{(k)}\left(\epsilon_{m\sigma}^{(k)} - \mu_{\sigma}^{(k)}\right); m\sigma = 1, \ldots, M_\sigma\right\}; \sigma = \alpha, \beta. \tag{63}$$

Practically, we can calculate only a set of 2^{n-1} principal internal energy functionals:

$$U_{UHFD}^{(k)}\left[\rho_\alpha^{(k)}, \rho_\beta^{(k)}\right] = T^{(k)}[\rho] + V_{ne}^{(k)}[\rho] + J^{(k)}[\rho] - K_{XC}^{(k)}\left[\rho_\alpha^{(k)}, \rho_\beta^{(k)}\right]; k = 1, \ldots, 2^{n-1}. \tag{64}$$

However, the origin of H_{ex} must be traced to a set of 2^{n-1} equality relationships between the principal exchange-correlation energy functional, $-K_{XC}^{(k)}\left[\rho_\alpha^{(k)}, \rho_\beta^{(k)}\right]$, and the projected value of the spin-dependent XC energy functional, i.e.

$$-K_{XC}^{(k)}\left[\rho_\alpha^{(k)}, \rho_\beta^{(k)}\right] = 2H_{ES}^{(k)}\left[\rho_\alpha^{(k)}, \rho_\beta^{(k)}\right] = 2\mathrm{Tr}_s\left(H_{ES}^{(k)}\left[\Gamma_\alpha^{(k)}, \Gamma_\beta^{(k)}\right]\right); k = 1, \ldots, 2^{n-1}. \tag{65}$$

We note that the kth projected value of $\left(S_i, S_j\right)$ onto the binary Hilbert space spanned by the kth GC ensemble $\mathbf{M}^{(k)}$ must depend not only on the kth principal exchange-correlation energy between *i*ES and *j*ES, but also on the polarized spins of bridging and non-bridging ligand atoms, *iLaj* ($j \neq i$) and *iLnb*, respectively, via the conservation law of each projected spin number of ${n_i}^{(k)}$ and ${n_j}^{(k)}$, defined by

$$n_i^{(k)} \equiv 2S_i\sigma_{z,i}^{(k)}; \sigma_{z,i}^{(k)} = \pm 1, k = 1, \ldots, 2^{n-1}; i = 1, \ldots, n. \tag{66}$$

$$\sum_{i=1}^{n} S_i\sigma_{z,i}^{(k)} \geq 0, k = 1, \ldots, 2^{n-1}. \tag{67}$$

Although the arrangement order of n ESA-codes $\left\{\sigma_{z,i}^{(k)} = 1 \text{ or } -1, i = 1-n\right\}$ leaves the choice of one's best depending on the arrangement order of n TM-cation spins $\{S_i, -1-n\}$, a non-negative sum rule of Eq. (67) must be satisfied owing to the time-reversal symmetry.

For the ith transition metal (TM) cations, its non-bridging ligand atoms $iLnb$ and its bridging ligand atoms $iLaj$ between the ith and jth TM-cations, this wave-packet spin projection may be entrusted to the respective spin operator by itself, that is $\boldsymbol{S}_i$, $\boldsymbol{s}_{iLnb}$ and $\boldsymbol{s}_{iLaj}$, by imposing each projection equation acting on a spin-dependent AO in a LCAO-NMO wave function, without change of the F-D distribution for the former two and with change to its half distribution for the latter one:

$$\boldsymbol{S}_i g_l^a(\boldsymbol{x}) = \delta_{a,i}\delta_{l,3d_i}\boldsymbol{S}_i g_{3d}^i(\boldsymbol{x}), \tag{68}$$

$$\boldsymbol{s}_{iLnb} g_l^a(\boldsymbol{x}) = \delta_{a\in iLnb}\delta_{l,2p}\boldsymbol{s}_{iLnb} g_{2p}^{iLnb}(\boldsymbol{x}), \tag{69}$$

$$\boldsymbol{s}_{iLaj} g_l^a(\boldsymbol{x}) = \delta_{a\in iLaj}\delta_{l,2p}\boldsymbol{s}_{iLaj} g_{2p}^{iLaj}(\boldsymbol{x}); f_{iLajm\sigma} = \frac{f_{im\sigma}}{\nu_{iLaj}} = \frac{f_{im\sigma}}{2}, \tag{70}$$

where we have introduced the share frequency $\nu_{iLajm\sigma}$ among n different subsets (in this case, it is 2), $\delta_{a\in A}$ = (1 for $a \in$ A; 0 for otherwise) and $\delta_{x,y}$ is Kronecker's δ.

Concomitantly, the kth GC ensemble $\mathbf{M}^{(k)}$ might be decomposed into n spin-dependent NMO-subsets associated with these elements, and the other spin-free subset as in Eq. (71), each $\mathbf{M}_i^{(k)}$ further decomposed into three components as in Eqs. (72) and (73), finally to define the ith ES in terms of two components in Eq. (74).

$$\mathbf{M}^{(k)} = \sum_{i=1}^{n}\mathbf{M}_i^{(k)} + \mathbf{M}'^{(k)}, \tag{71}$$

$$\mathbf{M}_i^{(k)} = \mathbf{M}_{id}^{(k)} + \mathbf{M}_{iL}^{(k)} + \mathbf{M}_{io}^{(k)}, \tag{72}$$

$$\mathbf{M}_{iL}^{(k)} = \sum_{j\neq i}^{n}\mathbf{M}_{iLaj}^{(k)} + \mathbf{M}_{iLnb}^{(k)}, \tag{73}$$

$$\mathbf{M}_{iES}^{(k)} = \mathbf{M}_{id}^{(k)} + \sum_{j\neq i}^{n}\mathbf{M}_{iLaj}^{(k)}, \tag{74}$$

where the ith subset in Eq. (72) consists of the subset $\mathbf{M}_{id}^{(k)}$ associated with (3d, 4 s, 4p)-electron AO's in the ith TM cation, the subset $\mathbf{M}_{iL}^{(k)}$ associated with 2p-valence electron AO's in the iLth assembly of ligand atoms, and the subset $\mathbf{M}_{io}^{(k)}$ associated with other doubly-occupied core-shell AO's in the ith TM cation. The iLth subset in Eq. (73) can be further decomposed into two kinds of ligand assembly: (1) a *thermal equipartition* half of the $iLaj$th assembly of bridging ligand atoms between the ith and jth TM cations, which can mediate the antiferromagnetic super-exchange coupling, and (2) the $iLnb$th assembly of non-bridging ligand atoms around the ith TM cation, which can be either paramagnetically or diamagnetically polarized depending on the ligand C/N/O atomic structure and hence control the ith ES magnitude via *the spin number* (n_i) *conservation law* governing the Mulliken atomic spin densities $\{M_a^{(k)}\}$ [14] of these magnetically-interacting atoms, finally to define the ith ES density $n_{iEF}^{(k)}$ so as to satisfy Eq. (76):

$$n_{iEF}^{(k)} \equiv M_{id}^{(k)} + \sum_{j\neq i}^{n} \sum_{iLaj} \frac{M_{iLaj}^{(k)} \vartheta\left(M_{iLaj}^{(k)} \sigma_{z,i}^{(k)}\right)}{\nu_{iLaj}^{(k)}} + \delta_{k>1} \sum_{j\neq i}^{n} \sum_{iLaj} \sigma_{z,i}^{(k)} \frac{M_{iLaj}^{(1)} \vartheta\left(-\sigma_{z,j}^{(1)} \sigma_{z,i}^{(1)}\right)}{\nu_{iLaj}^{(1)}}; \quad (75)$$

$$n_{iLnb}^{(k)} \equiv \sum_{iLnb'} M_{iLnb'}^{(k)}; n_{iEF}^{(k)} + n_{iLnb}^{(k)} = n_i.k = 1(\mathrm{F}), 2(\mathrm{AF}), \ldots, 2^{n-1}(\mathrm{AF}). \quad (76)$$

Here, ϑ(sign1*sign2) is the Heaviside step function, and $\nu_{iLaj}^{(k)}$ is the frequency of the *iLaj* spin density shared by plural ESs with the same sign, which can be calculated as

$$\nu_{iLaj}^{(k)} = 1 + \vartheta\left(\sigma_{z,i}^{(k)} \sigma_j^{(k)}\right) \text{ for all } k\text{'s}, \quad (77)$$

(CORRIGENDAM: Eq. (5.6c) in [13] should be replaced by Eq. (77)).

Decomposition into these NMO-subsets allows us to provide a noble systematic and quantitative method to derive a set of the expected values of H_{ex} $\{<H_{ex}>^{(k)}; k = 1, \ldots, 2^{n-1}\}$ from the spin-dependent XC energy functional in Eq. (65), as follows:

The kth spin density matrix for Eq. (42) can be decomposed into

$$Q^{(k)}(\boldsymbol{r},\boldsymbol{r}') = \sum_{\sigma=\alpha}^{\beta} (-1)^{\sigma} \sum_{m\sigma}^{{}^{\sigma}\mathbf{M}^{(k)}} f_{m\sigma} \Phi_{m\sigma}(\boldsymbol{r}) \Phi_{m\sigma}(\boldsymbol{r})^{*}; (-1)^{\alpha/\beta} = \pm 1, \quad (78)$$

$$= \sum_{i=1}^{n} \left[Q_{iES}^{(k)}(\boldsymbol{r},\boldsymbol{r}') + Q_{iLnb}^{(k)}(\boldsymbol{r},\boldsymbol{r}') + Q_{io}^{(k)}(\boldsymbol{r},\boldsymbol{r}')\right] + Q'^{(k)}(\boldsymbol{r},\boldsymbol{r}'), \quad (79)$$

$$= \sum_{i=1}^{n} \left[Q_{iES}^{(k)}(\boldsymbol{r},\boldsymbol{r}') + Q_{iLnb}^{(k)}(\boldsymbol{r},\boldsymbol{r}')\right] + \sum_{i=1}^{n} \left[\rho_{io,\alpha}^{(k)}(\boldsymbol{r},\boldsymbol{r}') - \rho_{io,\beta}^{(k)}(\boldsymbol{r},\boldsymbol{r}')\right] + \rho'^{(k)}_{\alpha}(\boldsymbol{r},\boldsymbol{r}') - \rho'^{(k)}_{\beta}(\boldsymbol{r},\boldsymbol{r}'), \quad (80)$$

in which Eq. (80) indicates that the first term will contribute to both intra-atomic and interatomic spin-density coupling energies generated by open 3d-shell electrons, as given by

$$2H_{ES1}^{(k)}\left[\rho_{\alpha}^{(k)}, \rho_{\beta}^{(k)}\right] \equiv 2E_{ES1}^{(k)} = -\sum_{i=1}^{n} J_{i,i}^{(k)} S_i(S_i+1), \quad (81)$$

$$J_{i,i}^{(k)} \equiv \frac{1}{2n_i^2} \iint \frac{d\boldsymbol{r}_1 d\boldsymbol{r}_2}{r_{12}} \left[\rho_{iES}^{(k)}(\boldsymbol{r}_1,\boldsymbol{r}_2) + \rho_{iLnb}^{(k)}(\boldsymbol{r}_1,\boldsymbol{r}_2)\right] \left[\rho_{iES}^{(k)}(\boldsymbol{r}_2,\boldsymbol{r}_1) + \sigma_{iLnb}^{(k)}(\boldsymbol{r}_2,\boldsymbol{r}_1)\right], \quad (82)$$

and

$$2H_{ES2}^{(k)}\left[\rho_{\alpha}^{(k)}, \rho_{\beta}^{(k)}\right] = 2E_{ES2}^{(k)} = \langle H_{ex}\rangle^{(k)} = -2\sum_{i<j}^{n} J_{i,j} \langle (\boldsymbol{S}_i, \boldsymbol{S}_j) \rangle^{(k)}, \quad (83)$$

$$\langle (\boldsymbol{S}_i, \boldsymbol{S}_j) \rangle^{(k)} = \mathrm{Tr}\left[(\boldsymbol{S}_i, \boldsymbol{S}_j) \Gamma_2^{(k)}(\boldsymbol{x}_1\boldsymbol{x}_2, \boldsymbol{x}_1\boldsymbol{x}_2)\right], \quad (84)$$

$$= \sum_{M_i=-S_i}^{S_i} \sum_{M_j=-S_j}^{S_j} \mathrm{Tr}\left[\langle M_i, M_j | (\boldsymbol{S}_i, \boldsymbol{S}_j) | M_i, M_j \rangle \times \langle M_i, M_j | \Gamma_{i,j}^{(k)}(\boldsymbol{x}_1\boldsymbol{x}_2, \boldsymbol{x}_1\boldsymbol{x}_2) | M_i, M_j \rangle\right], \quad (85)$$

with

$$\Gamma_{i,j}^{(k)}(\boldsymbol{x}_1\boldsymbol{x}_2,\boldsymbol{x}_1\boldsymbol{x}_2)=\frac{2}{2}\left[\Gamma_{iES}^{(k)}(\boldsymbol{x}_1,\boldsymbol{x}_1)\Gamma_{jES}^{(k)}(\boldsymbol{x}_2,\boldsymbol{x}_2)-\Gamma_{iES}^{(k)}(\boldsymbol{x}_1,\boldsymbol{x}_2)\Gamma_{jES}^{(k)}(\boldsymbol{x}_2,\boldsymbol{x}_1)\right], \tag{86}$$

respectively, and the second and third terms to a major H_{ex}-less. component in the principal XC energy, as given by

$$-K_{XC-H_{ex}}^{(k)}\left[\rho_\alpha^{(k)},\rho_\beta^{(k)}\right]=2H_{ESO}^{(k)}\left[\rho_{ESO,\alpha}^{(k)},\rho_{ESO,\beta}^{(k)}\right]=2E_{ESO}^{(k)}, \tag{87}$$

$$\begin{aligned}=-\frac{1}{2}\iint\frac{d\boldsymbol{r}_1d\boldsymbol{r}_2}{r_{12}}&\left[\rho^{(k)}(\boldsymbol{r}_1,\boldsymbol{r}_2)-\sum_{i=1}^{n}\left\{\rho_{iES}^{(k)}(\boldsymbol{r}_1,\boldsymbol{r}_2)+\rho_{iLnb}^{(k)}(\boldsymbol{r}_1,\boldsymbol{r}_2)\right\}\right]\\ \times&\left[\rho^{(k)}(\boldsymbol{r}_2,\boldsymbol{r}_1)-\sum_{i=1}^{n}\left\{\rho_{iES}^{(k)}(\boldsymbol{r}_2,\boldsymbol{r}_1)+\rho_{iLnb}^{(k)}(\boldsymbol{r}_2,\boldsymbol{r}_1)\right\}\right];\end{aligned} \tag{88}$$

To calculate Eq. (84), we need the total spin operator given by

$$\boldsymbol{S}_{tot}=\sum_{i=1}^{n}\left(\boldsymbol{S}_i+\sum_{iLaj}\boldsymbol{s}_{iLaj}+\sum_{iLnb}\boldsymbol{s}_{iLnb}\right), \tag{89}$$

The ith ES spin operator, $\boldsymbol{S}_i$, is considered to turn around between up and down states, $|S_i>$ and $|-S_i>$, in the binary Hilbert space spanned by two rules:

$$\sum_{M_i=-S_i}^{S_i}|M_i\rangle\langle M_i|=\mathbf{1}_i; i=1,\dots,n, \tag{90}$$

and

$$\langle\alpha_i|M_i\rangle=\frac{\delta_{M_i,S_i}}{\sqrt{|n_i|}},\langle\beta|M_i\rangle=\frac{\delta_{M_i,-S_i}}{\sqrt{|n_i|}}; \tag{91}$$

while the $iLaj$th and $iLnb$th ES spin operators, $\boldsymbol{s}_{iLaj}$ and $\boldsymbol{s}_{iLnb}$, are assumed to automatically respond to the up or down state of $\boldsymbol{S}_i$. Then, the expected value of the z-component of $\mathbf{S}_{tot}$ in the kth ESA state is given by

$$\langle S_{tot,z}\rangle^{(k)}=\sum_{i=1}^{n}\mathrm{Tr}\left\{S_{i,z}\left[\Gamma_{iES}^{(k)}(\boldsymbol{x},\boldsymbol{x})+\Gamma_{iLnb}^{(k)}(\boldsymbol{x},\boldsymbol{x})\right]\right\}, \tag{92}$$

$$=\sum_{i=1}^{n}S_i\sigma_{z,i}^{(k)}\left(P_{iES}^{(k)}+P_{iLnb}^{(k)}\right)=\sum_{i=1}^{n}S_i\sigma_{z,i}^{(k)}, \tag{93}$$

$$\Gamma_{iES}^{(k)}(\boldsymbol{x},\boldsymbol{x}')=\Gamma_{id}^{(k)}(\boldsymbol{x},\boldsymbol{x}')+\sum_{j\neq i}^{n}\sum_{iLaj'}\Gamma_{iLaj'}^{(k)}(\boldsymbol{x},\boldsymbol{x}'), \tag{94}$$

$$P_{iES}^{(k)}=\frac{n_{iES}^{(k)}}{n_i}=\frac{1}{n_i}\int d\boldsymbol{r}Q_{iES}^{(k)}(\boldsymbol{r},\boldsymbol{r}), \tag{95}$$

$$P_{iLnb}^{(k)}=\frac{n_{iLnb}^{(k)}}{n_i}=\frac{1}{n_i}\int d\boldsymbol{r}Q_{iLnb}^{(k)}(\boldsymbol{r},\boldsymbol{r}), \tag{96}$$

$$P_{iES}^{(k)} + P_{iLnb}^{(k)} = \left[n_{iES}^{(k)} + n_{iLnb}^{(k)}\right]/n_i = 1. \tag{97}$$

It is important to remind that the last additional term in Eq. (75) in the antiferromagnet *kAF*th ESA state was absolutely required for a systematic better agreement with experimental $J_{i,j}$ indicating that a large-positive $M_{iLaj}^{(1F)}$ density in the *1F*th-ESA state can be divided into two half densities which will be reversed to be distributed with phase matching to an antiferromagnetic pair of $n_{iES}^{(kAF)}$ and $n_{jES}^{(kAF)}$ in the *kAF*th ESA state, as given by a programmatic equation [13],

$$M_{iLaj}^{(k)} \Leftarrow \frac{1}{2}\sigma_{z,i}^{(k)} M_{iLaj}^{(1)} \vartheta\left(-\sigma_{z,i}^{(k)}\sigma_{z,j}^{(k)}\right) + M_{iLaj}^{(k)}. \tag{98}$$

Now, substituting Eq. (86) into Eq. (85) and taking the traces over spin-orbital coordinates with use of Eqs. (90), (91), (95) and (96), we obtain a noble formula

$$\left\langle \left(\mathbf{S}_i, \mathbf{S}_j\right)\right\rangle^{(k)} = n_{iES}^{(k)} n_{jES}^{(k)}\left(1 - S_{iES,jES}^{(k)}\right), \tag{99}$$

$$S_{iES,jES}^{(k)} \equiv \frac{\mathrm{Tr}\left[\Gamma_{iES}^{(k)}(\mathbf{x}_1, \mathbf{x}_2)\Gamma_{jES}^{(k)}(\mathbf{x}_2, \mathbf{x}_1)\right]}{\mathrm{Tr}\left[\Gamma_{iES}^{(k)}(\mathbf{x}_1, \mathbf{x}_1)\right]\mathrm{Tr}\left[\Gamma_{jES}^{(k)}(\mathbf{x}_1, \mathbf{x}_1)\right]}, k = 1, \dots, 2^{n-1}. \tag{100}$$

Here, $\{S_{iES,jES}^{(k)}\}$ represents a set of the Exchange-Correlation vs. Classical Coulomb Density Overlap Integral (XC/CC-DOI) ratios. Although it appears almost impossible to directly calculate a set of $2^{n-2}n(n\text{-}1)$ XC/CC-DOI ratios, we could find out a reasonable solution of Eq. (103) by imposing 2^{n-1} equations to eliminate all the residues $\{\Delta^2\langle \mathbf{S}_{tot}^2\rangle^{(k)}\}$ from a set of the expected values of $\{\langle \mathbf{S}_{tot}^2\rangle^{(k)}\}$, which are given by

$$\langle \mathbf{S}_{tot}^2\rangle^{(k)} = \langle S_{tot,z}\rangle^{(k)}\left[\langle S_{tot,z}\rangle^{(k)} + 1\right] + \sum_{i=1}^{n} S_i\left(1 - \sigma_{z,i}^{(k)}\right) + \Delta^2\langle \mathbf{S}_{tot}^2\rangle^{(k)}, \tag{101}$$

$$\Delta^2\langle \mathbf{S}_{tot}^2\rangle^{(k)} = \sum_{i=1}^{n} S_i^2\left[1 - \left(P_{iES}^{(k)}\right)^2\right] - \frac{1}{2}\sum_{i<j} n_{iES}^{(k)} n_{jES}^{(k)} S_{iES,jES}^{(k)} = 0, \tag{102}$$

$$S_{iES,jES}^{(k)} = \frac{4}{n(n-1)}\left\{\sum_{i'=1}^{n} S_{i'}^2\left[1 - \left(n_{i'ES}^{(k)}/2S_{i'}^{(k)}\right)^2\right]\right\}\left(n_{iES}^{(k)} n_{jES}^{(k)}\right)^{-1}; k = 1, \dots, 2^{n-1}, \tag{103}$$

Thus, we could derive the MGC-set of the internal energy functionals taking each different decomposition form from Eq. (64) involving the projected Heisenberg spin Hamiltonian:

$$\langle H_{ex}\rangle^{(k)} = -\frac{1}{2}\sum_{i<j}^{n} J_{i,j}\left(1 - S_{iES,jES}^{(k)}\right) n_{iES}^{(k)} n_{jES}^{(k)}, k = 1, \dots, 2^{n-1}, \tag{104}$$

$$U_{UHFD-H_{ex}}^{(k)}[\rho] = T^{(k)}[\rho] + V_{en}^{(k)}[\rho] + J^{(k)}[\rho] - K_{XC-Hex}^{(k)}\left[\rho - \sum_{i=1}^{n}\{\rho_{iES} + \rho_{iLnb}\}\right], \tag{105}$$

$$U_{UHFD}^{(k)}\left[\rho_\alpha,\rho_\beta\right]=U_{UHFD-H_{ex}}^{(k)}[\rho]+\langle H_{ex}\rangle^{(k)}. \tag{106}$$

Notably, one may expect that the H_{ex}-less internal energy function defined by Eq. (106) will become almost constant over 2^{n-1} ESA states, owing to the sum of four different components with each having a weak k-dependency. If this is the case, one can utilize Eqs. (104) and (106) to determine the only unknown set of mean ES-exchange coupling constants, $\{J_{i,j}; i(j>i)=1,\dots,n\}$, since $n2^{n-1}$ effective spin densities $\{n_{iES}{}^{(k)}\}$, $2^{n-2}n(n\text{-}1)$ XC/CC-DOI ratios defined by Eq. (103) and $(2^{n-1}\text{-}1)$ energy-difference equations (107) could be quantitatively calculated using UB3LYP/PBS(ε6)/lacvp** method:

$$\Delta U_{UHFD}^{(k)}\left[\rho_\alpha,\rho_\beta\right]\equiv U_{UHFD}^{(k)}\left[\rho_\alpha,\rho_\beta\right]-U_{UHFD}^{(1)}\left[\rho_\alpha,\rho_\beta\right]\cong\langle H_{ex}\rangle^{(k)}-\langle H_{ex}\rangle^{(1)};k=2,\dots,2n-1, \tag{107}$$

so that the transformed equations can be written in regular matric form:

$$\boldsymbol{A}^T\boldsymbol{A}\boldsymbol{X}=\boldsymbol{A}^T\boldsymbol{B}, \tag{108}$$

$$\boldsymbol{A}=-\frac{1}{2}\begin{bmatrix}\Delta\left[\left(1-S_{1ES,2ES}^{(2)}\right)n_{1ES}^{(2)}n_{2ES}^{(2)}\right] & \cdots & \Delta\left[\left(1-S_{(n-1)ES,nES}^{(2)}\right)n_{(n-1)ES}^{(2)}n_{nES}^{(2)}\right]\\ \vdots & \ddots & \vdots\\ \Delta\left[\left(1-S_{(n-19ES,nES}^{(2^{n-1})}\right)n_{(n-1)ES}^{(2^{n-1})}n_{nES}^{(2^{n-1})}\right] & \cdots & \Delta\left[\left(1-S_{(n-1)ES,nES}^{(2^{n-1})}\right)n_{(n-1)ES}^{(2^{n-1})}n_{nES}^{(2^{n-1})}\right]\end{bmatrix}, \tag{109}$$

$$\boldsymbol{B}=(B_2,\cdots,B_{2^{n-1}}),B_k=\Delta U_{UHFD}^{(k)}-\Delta U_{UHFD-H_{ex}}^{(k)};k=2,\dots,2^{n-1}, \tag{110}$$

$$\boldsymbol{X}=(X_1,\cdots,X_{n(n-1)/2})\equiv(X_{ij});X_{ij}=J_{i,j}, \tag{111}$$

where $\boldsymbol{A}^T$ is the transpose of $\boldsymbol{A}$. Thus, we get a unique solution:

$$J_{i,j}=\left[\left(\boldsymbol{A}^T\boldsymbol{A}\right)^{-1}\boldsymbol{A}^T\boldsymbol{B}\right]_{ij}. \tag{112}$$

For n = 2, Eq. (112) reduces to

$$J_{1,2}\cong\frac{2\Delta U_{UHFD}^{(2)}}{\left(1-S_{1ES,2ES}^{(1)}\right)n_{1ES}^{(1)}n_{2ES}^{(1)}-\left(1-S_{1ES,2ES}^{(2)}\right)n_{1ES}^{(2)}n_{2ES}^{(2)}}. \tag{113}$$

In **Table 1**, we show again the results of benchmark-test calculations of the ES-exchange coupling constants $J_{1,2}$, designated $(J_{1,2/\mathbf{a}}^{(m)},J_{1,2/\mathbf{c}}^{(m)},\dots,J_{1,2/\mathbf{k}}^{(m)})$, for 10 biomimetic binuclear Cu, Mn and Fe complexes, named (**a**, **c**, ..., **k**), were made using 13 conventional mXC/PBS/lacvp** method (m = 4 ~ 16) in place of the present MGC-SDFT-UHFD (≡0XC) method, which is unfortunately not yet implemented. These data sets were compared with the observed values, named $\left(J_{1,2/\mathbf{a}}^{exp},J_{1,2/\mathbf{c}}^{exp},\dots,J_{1,2/\mathbf{k}}^{exp}\right)$, to show all the excellent quantitative agreements between the theoretical values $\left(J_{1,2/\mathbf{a}}^{(4)},J_{1,2/\mathbf{c}}^{(4)},\dots,J_{1,2/\mathbf{k}}^{(4)}\right)$ and the experimental values mentioned above only by the standard B3LYP (≡4XC) method [13]. Here, we raise two possible explanations for the best performance by the B3LYP hybrid XC energy functional; (1) the best atomic structure of each TM-dimeric complex could be obtained by further geometry-optimization near the observed XRD structure by the B3LYP/PBS(ε)/lacvp** method [15, 16] with the dielectric constant ε of the solvent being chosen the best one from 5, 10, 20, and 40 [13]; (2) as *an ideally-good balance between the exchange and correlation energy* in the UHFD approximation is considered to be a key factor, B3LYP/PBS(ε)/ lacvp** method may satisfy this condition most closely.

2.4 Two most stable isomers of the $S_1^{(0)}$ state Mn_4CaO_5 clusters: Identified by two EPR signals

We have recently applied the UB3LYP/PBS(ε)/lacvp** method in place of the UHFD/ PBS(ε)/lacvp** method to all the water-splitting and oxygen-evolving reactions catalyzed by the Mn_4CaO_5 cluster in photosystem II (PSII). The electron-abstracting and proton-releasing reactions from the so-called oxygen-evolving complex (OEC) are considered to occur serially via five redox states, called Kok's S_i-states (i = 0, 1, ..., 4), where S_1 is the dark-stable state, and S_4 spontaneously decays to the initial S_0-state after releasing two protons and evolving dioxygen: the generalized reaction schemes are symbolically given by

$$S_1^{(0)} \overset{-e}{\to} S_2^{(+1)} \overset{-e-H^+}{\to\to} S_3^{(+1)} \overset{-e-2H^+}{\to\to} S_4^{(0)} \overset{+2H_2O-O_2}{\to\to} S_0^{(0)} \overset{-e-H^+}{\to\to} \left\{S_1^{(0)} \circlearrowleft\right\}, \tag{114}$$

where the figure k in the superfix parentheses of $S_i^{(k)}$ represents a formal charge of the ith OEC, −e above an arrow (→) indicate one electron transfer from OEC to $P680^{(+)}$, an oxidized PSII reaction center intermittingly generated by every ~10 μs light-pulse, $-H^+$above an arrow (→) does a proton released into aqueous phase, and the symbols +(−) indicate to go out(in) of OEC, respectively. Among many controversial problems remained to be elucidated, we here take up the molecular structure of the $S_1^{(0)}$-state Mn_4CaO_5 cluster, that is not yet established because the experimental data from XFEL, EPR and EXAFS spectroscopies appear to be apparently inconsistent if these are assumed to have been observed from the same $S_1^{(0)}$-state. Although we can'nt exclude the possibility that the XFEL model [21] may reflect a photo-reduced $S_0^{(-1)}$ state of the $S_1^{(0)}$-state Mn_4CaO_5 cluster, we have no reason to doubt the fact that two kinds of broad g4.8 and g12-multiline EPR signals were observed from the $S_1^{(0)}$-state samples of cyanobacteria [18, 19] and spinach [20], respectively, which must have slightly different structures due to the different peripheral proteins between them. Indeed, we could prove that these EPR signals are attributable to two different structural isomers, named S_{1A} and S_{1B} in [23], which coexist with quasi-degenerate lowest energies in the respective $S_1^{(0)}$-state. Two papers substantiating these ideas will be submitted for publication near future.

2.5 Superconductivity enhanced by the secondary correlation interaction in metals

It is well known that many materials become superconducting (S-phase) at lower temperatures than the critical temperature Tc where each system makes the transition from the normal metallic phase (N-phase). This phenomenon has been explained in terms of the Bardeen-Cooper-Schrieffer model [23, 24] combining the Fröhlich electron-lattice attractive interaction model [25] and the Bogoliubov Cooper-pairing model [26]. The highest T_c that had been achieved on 2015 is the sulfur hydride system at 203 K at high pressure (155 GP), identified from the observed magnetization vs. θ/k_B stepping-curve [26]. The observed H/D isotope effect on the down-shift & down-size of this curve appears to be consistent with the BCS model. Drozdov et al. raised three conditions required for such much higher-T_c than those of normal metals: (1) higher-frequency-phonon, (2) stronger electron-phonon coupling, and (3) a higher-density of Cooper pairing states [27]. At least the former two conditions could in principle be fulfilled for metallic and covalent compounds dominated by hydrogen. But notably the BCS model contains a serious deficiency that it is based on the free electron model but not on the Bloch-electron model depending on any approximation of self-consistent exchange-correlation potential, so that it is forced to take account only the screened Coulomb repulsive force between conductive electrons, as given by a Fourie transform,

$$\lim_{\kappa\to 0^+}\frac{e^2}{r}e^{-\kappa r} = \int V_{sc}(\boldsymbol{q})e^{i\boldsymbol{q}\bullet\boldsymbol{r}}d\mathbf{q};\, V_{sc}(\boldsymbol{q}) = \lim_{\kappa\to 0^+}\frac{4\pi e^2}{q^2+\kappa^2}, \tag{115}$$

where κ is called "Thomas-Fermi wave number" and the limit of $\kappa \to 0^+$ implies a bare-Coulomb interaction. In order to treat such a many-body effect for Bloch-electrons near the Femi-level μ_F, we should adopt the grand potential Ω as the more appropriate thermodynamic free energy than the internal energy U, as given by

$$\Omega = -pV = U - \mu_F\overline{N} - \theta\Sigma. \tag{116}$$

Although here we assume that the decrease of entropy Σ upon the N-to-S phase transition may be relatively much smaller than the decrease of $U - \mu_F\overline{N}$ at least at low temperatures. This choice of Ω is also consistent with the fact that the higher pressure appears to be directly correlated with higher-T_c superconductors [27].

In contrast to the conventional idea of repulsive Coulomb force, the principal exchange-correlation energy Hamiltonian $\widehat{H}^0_{UHFD}$ of Eq. (57) is already incorporated in the present GC-UHFD-SDFT theory to define a binary set of Bloch eigenstates occupied with the Fermi-Dirac distribution $f(\epsilon_{\boldsymbol{k}} - \mu_F)$, designated ${}^{\sigma}_{B}\mathbf{M}(\sigma=\alpha, \beta)$, and there remains only the secondary correlation energy part of Eq. (60) to be regarded as giving rise to the attractive exchange force between opposite-spin itinerant Bloch-electrons. Hence, neglecting the entropic term, we need only to treat a small part of the grand potential of Eqs. (116) alone, that can be expanded by the perturbation theory:

$${}_B\Omega_{UHFD} = {}_B\Omega^0_{UHFD} + {}_B\Omega^1_{UHFD} + {}_B\Omega^2_{UHFD} + \cdots, \tag{117}$$

$${}_B\widehat{\Omega}^0_{UHFD} \cong {}_B\widehat{H}^0_{UHFD} - \mu_F\sum_{\boldsymbol{k}\sigma}^{{}^{\sigma}_{B}\mathbf{M}} a^\dagger_{\boldsymbol{k}\sigma}a_{\boldsymbol{k}\sigma} = \sum_{\sigma=\alpha,\beta}\sum_{\boldsymbol{k}\sigma}^{{}^{\sigma}_{B}\mathbf{M}}[\epsilon_{\boldsymbol{k}\sigma} - \mu_F]a^\dagger_{\boldsymbol{k}\sigma}a_{\boldsymbol{k}\sigma}, \tag{118}$$

$${}_B\widehat{\Omega}^1_{UHFD} = \sum_{\sigma=\alpha,\beta}\sum_{\boldsymbol{k}\sigma,\boldsymbol{k}'\sigma}^{{}^{\sigma}_{B}\mathbf{M}} V^{\boldsymbol{k}\sigma,\boldsymbol{k}'\overline{\sigma}}_{UHFD} V_{\boldsymbol{k}\,\sigma,\boldsymbol{k}'\overline{\sigma}} a^\dagger_{\boldsymbol{k}\,\sigma}a^\dagger_{\boldsymbol{k}'\overline{\sigma}}a_{\boldsymbol{k}'\overline{\sigma}}a_{\boldsymbol{k}\,\sigma}, \tag{119}$$

$$\begin{aligned} V^{\boldsymbol{k}\sigma,\boldsymbol{k}'\overline{\sigma}}_{UHFD} &\equiv -e^2\iint\frac{dr_1dr_2}{2r_{12}}\Phi^*_{\boldsymbol{k}\,\sigma}(\boldsymbol{r}_2)\Phi^*_{\boldsymbol{k}'\overline{\sigma}}(\boldsymbol{r}_1)\Phi_{\boldsymbol{k}'\overline{\sigma}}(\boldsymbol{r}_2)\Phi_{\boldsymbol{k}\,\sigma}(\boldsymbol{r}_1) \\ &\quad\times(\sigma_{+,\boldsymbol{k}\,\sigma,}\sigma_{-,\boldsymbol{k}'\overline{\sigma}} + \sigma_{-,\boldsymbol{k}\,\sigma}\sigma_{+,\boldsymbol{k}'\overline{\sigma}}) \end{aligned}, \tag{120}$$

$$\approx -\lim_{\kappa\to 0^+}\frac{4\pi e^2V}{\left|\boldsymbol{k}-\boldsymbol{k}'\right|^2+\kappa^2}(\sigma_{+,\boldsymbol{k}\,\sigma,}\sigma_{-,\boldsymbol{k}'\overline{\sigma}} + \sigma_{-,\boldsymbol{k}\,\sigma}\sigma_{+,\boldsymbol{k}'\overline{\sigma}}), \tag{121}$$

where we put $\overline{\alpha} = \beta$ and $\overline{\beta} = \alpha$, and used the plane-wave approximation for the conductive Bloch-wave functions: $\Phi_{k\sigma}(r_2) \approx \exp\,[i\boldsymbol{k}\bullet\boldsymbol{r}_2]$, $\Phi^*_{k\,\sigma}(r_2) \approx \exp\,[-i\boldsymbol{k}\bullet\boldsymbol{r}_2]$ et c. to make the double integrations on the coordinate vectors $\boldsymbol{r}_1$ and $\boldsymbol{r}_2$ after transformed into $\boldsymbol{R}_{12} = (\boldsymbol{r}_1 + \boldsymbol{r}_2)/2$ and $\boldsymbol{r}_{12} = \boldsymbol{r}_2 - \boldsymbol{r}_1$, together with introducing the screened-Coulomb damping factor $\exp.(-\kappa r_{12})$. Note that the volume of the system V appears from the integral on the center-of-mass coordinate $\boldsymbol{R}_{12}$.

Notably, this spin-dependent first-order perturbation of Eq. (120) is off-diagonal in the binary Hilbert space, so that it can'nt contribute to the renormalized eigenstates (S-state) through any odd-number order of perturbation term. Then, the predominant contribution could arise from the second order perturbation term as given by

$$ {}_B\Omega^2_{UHFD} = \sum_{\sigma=\alpha,\beta} \sum_{\boldsymbol{k}_{1\sigma}}^{{}^{\sigma}_{B}\mathbf{M}} \sum_{\boldsymbol{k}'_{1\overline{\sigma}}}^{{}^{\overline{\sigma}}_{B}\mathbf{M}} \sum_{\boldsymbol{k}_{2\sigma}}^{{}^{\sigma}_{B}\mathbf{M}} \sum_{\boldsymbol{k}'_{2\overline{\sigma}}}^{{}^{\overline{\sigma}}_{B}\mathbf{M}} \frac{V^{\boldsymbol{k}_1,\boldsymbol{k}'_1}_{UHFD} \left[V^{\boldsymbol{k}_2,\boldsymbol{k}'_2}_{UHFD}\right]^*}{\epsilon_{\boldsymbol{k}_{1\sigma}} + \epsilon_{\boldsymbol{k}'_{1\overline{\sigma}}} - \epsilon_{\boldsymbol{k}_{2\sigma}} - \epsilon_{\boldsymbol{k}'_{2\overline{\sigma}}}} \tag{122} $$

$$ \times f\left(\epsilon_{\boldsymbol{k}_{1\sigma}} - \mu\right) f\left(\epsilon_{\boldsymbol{k}'_{1\overline{\sigma}}} - \mu\right)\left[1 - f\left(\epsilon_{\boldsymbol{k}_{2\sigma}} - \mu\right)\right]\left[1 - f\left(\epsilon_{\boldsymbol{k}'_{2\overline{\sigma}}} - \mu\right)\right], $$

$$ V^{\boldsymbol{k},\boldsymbol{k}'}_{UHFD} = -\lim_{\kappa\to 0^+} \frac{4\pi e^2 V}{\left|\boldsymbol{k} - \boldsymbol{k}'\right|^2 + \kappa^2}. \tag{123} $$

Significantly in the second-quantization representation this term may be transformed into

$$ {}_B\widehat{\Omega}^2_{UHFD} = \sum_{\sigma=\alpha,\beta} \sum_{\boldsymbol{k}_{1\sigma}}^{{}^{\sigma}_{B}\mathbf{M}} \sum_{\boldsymbol{k}'_{1\overline{\sigma}}}^{{}^{\overline{\sigma}}_{B}\mathbf{M}} \sum_{\boldsymbol{k}_{2\sigma}}^{{}^{\sigma}_{B}\mathbf{M}} \sum_{\boldsymbol{k}'_{2\overline{\sigma}}}^{{}^{\overline{\sigma}}_{B}\mathbf{M}} V^{\boldsymbol{k}_1,\boldsymbol{k}'_1;\,\boldsymbol{k}_2,\boldsymbol{k}'_2}_{UHFD} \tag{124} $$

$$ \times \widehat{a}_{\boldsymbol{k}'_{2\overline{\sigma}}} \widehat{a}^\dagger_{\boldsymbol{k}'_{2\overline{\sigma}}} \widehat{a}_{\boldsymbol{k}_{2\sigma}} \widehat{a}^\dagger_{\boldsymbol{k}_{2\sigma}} \widehat{a}^\dagger_{\boldsymbol{k}'_{1\overline{\sigma}}} \widehat{a}_{\boldsymbol{k}'_{1\overline{\sigma}}} \widehat{a}^\dagger_{\boldsymbol{k}_{1\sigma}} \widehat{a}_{\boldsymbol{k}_{1\sigma}}, $$

$$ V^{\boldsymbol{k}_1,\boldsymbol{k}'_1;\,\boldsymbol{k}_2,\boldsymbol{k}'_2}_{UHFD} = \lim_{\kappa\to 0^+} \frac{\left(4\pi e^2 V\right)^2}{\left(\left|\boldsymbol{k}_1 - \boldsymbol{k}'_1\right|^2 + \kappa^2\right)\left(\left|\boldsymbol{k}_2 - \boldsymbol{k}'_2\right|^2 + \kappa^2\right)\left(\epsilon_{\boldsymbol{k}_1} + \epsilon_{\boldsymbol{k}'_1} - \epsilon_{\boldsymbol{k}_2} - \epsilon_{\boldsymbol{k}'_2}\right)} < 0. \tag{125} $$

The first point to notice is that the second-order perturbation Eq. (122) might be too complicated but could generate the attractive interaction between two Cooper-pair particles if it be approximated by the appropriate form (simply putting $\boldsymbol{k}'_{i\beta} = -\boldsymbol{k}_{i\beta}$; $i = 1, 2$ and multiplying twice the state number in each spherical-shell volume, $4\pi k_F^3 \omega_D/\mu_F$, as given by $N(k_F) \approx 4\pi k_F^3 \omega_D/\mu_F\ (2\pi)^3 V = k_F^3 \omega_D / 2\pi^2 V \mu_F$):

$$ {}_B\widehat{\Omega}^2_{UHFD} \approx \sum_{\sigma=\alpha,\beta} \sum_{\boldsymbol{k}_{1\sigma}}^{{}^{\sigma}_{B}\mathbf{M}} \sum_{\boldsymbol{k}_{2\sigma}}^{{}^{\sigma}_{B}\mathbf{M}} V^{\boldsymbol{k}_1,-\boldsymbol{k}_1;\,\boldsymbol{k}_2,-\boldsymbol{k}_2}_{UHFD} \widehat{a}_{-\boldsymbol{k}_{2\overline{\sigma}}} \widehat{a}^\dagger_{-\boldsymbol{k}_{2\overline{\sigma}}} \widehat{a}_{\boldsymbol{k}_{2\sigma}} \widehat{a}^\dagger_{\boldsymbol{k}_{2\sigma}} \widehat{a}^\dagger_{-\boldsymbol{k}_{1\overline{\sigma}}} \widehat{a}_{-\boldsymbol{k}_{1\overline{\sigma}}} \widehat{a}^\dagger_{\boldsymbol{k}_{1\sigma}} \widehat{a}_{\boldsymbol{k}_{1\sigma}}, \tag{126} $$

$$ V^{\boldsymbol{k}_1,-\boldsymbol{k}_1;\,\boldsymbol{k}_2,-\boldsymbol{k}_2}_{UHFD} \approx \lim_{\kappa\to 0^+} \frac{\left(4\pi e^2\right)^2 \left(k_F^3 \omega_D / 2\pi^2 \mu_F\right)^2}{2\left(4k_1^2 + \kappa^2\right)\left(4k_2^2 + \kappa^2\right)\left(\epsilon_{\boldsymbol{k}_1} - \epsilon_{\boldsymbol{k}_2}\right)} < 0; \tag{127} $$

which is an attractive potential under the BCS restrictions:

$$ -\omega_D < \epsilon_{\boldsymbol{k}_1} - \mu_F < 0 < \epsilon_{\boldsymbol{k}_2} - \mu_F < \omega_D, \tag{128} $$

where ω_D is the Debye frequency, and note that the matrix element Eq. (126) does not contain V and it is proportional to $\omega_D{}^2/(\epsilon_{k_2} - \epsilon_{k_1})$, which diverges as $(\epsilon_{k_2} - \epsilon_{k_1}) \to 0$ and hence may not be approximated as a constant. Examination of this singularity problem must be postponed in future, because of the page limitation.

Up to the present stage, however, we find out that in the principal GC-SDFT-UHFD method the remained secondary correlation interaction between Bloch-electrons near the Fermi-surface could generate an additional attractive force to promote the Cooper-pairing superconductivity by increasing not only the concentration of Cooper-pair particles but also the energy gap at the Fermi level.

3. Conclusion

In this chapter, we have reviewed the MGC-SDFT-UHFD method proposed in [13] in order to advance beyond the conventional KS-DFT-UHF method. We need more clearly to explain why the KS-formalism must be regarded as incomplete, because it is a kind of double standard or hybrid theory based the quantum-mechanical rule in closed system and the thermodynamic rule in open system, as clearly seen from their use of two distinct variation-principal equations. This inconsistent theory results in two problematic notions, (1) "eternally-unknown correlation energy functional" including a separated part of kinetic energy, and (2) a set of mutually interacting LCAO-MO quasi-particles.

Here, we have widely proposed a thermodynamic alternative to derive the principal internal energy functional, which has been required to define the self-consistent one-body potential in the Schrödinger equation yielding the ultimate ground and excited states, further which have been required multiple grand canonical ensembles to properly describe all kinds of spin-dependent systems, like the paramagnetic properties of the water-splitting Mn_4CaO_5-cluster in photosystem II. This one-body quasi-particle world picture has been completed by our two revolutionary discoveries of the *principal exchange-correlation energy functional,* that is, a non-local exchange-correlation interaction, and a complete set of self-consistent *LCAO-NMOs*, which extensively span all the energy levels below dissociation limit (called the work function W) with the Fermi-Dirac distribution.

Significantly, we have presented in Sections 2.3 and 2.4 two experimental evidences directly supporting the quantitative and systematic aspects of the MGC-SDFT-UHFD method, and in Section 2.5 one more evidence indirectly supporting this UHFD decoupling scheme retaining the only *secondary correlation energy functional*, which spin-dependent interaction between Bloch-electrons can promote Cooper-pairings of Bloch-electrons near Fermi-level in superconductor, provided that their eigen states might be exactly determined by the MGC-SDFT-UHFD method under the crystalline periodic conditions. This implies that the Bloch-electrons near the Fermi surface are unstable in the normal phase and hence tend to make the phase transition to the superconducting phase. Further, this provides an additional mechanism for the high-temperature superconductivity. It is further emphasized that the MGC-SDFT-UHFD/PSB(ε)/lacvp** method can help meet the demand for an eagerly awaited, first principle, quantitative, and practical method to elucidate the enzymatic function of paramagnetic Mn_4CaO_x clusters in a series of water-splitting and oxygen-evolving reactions in PSII. Moreover, the present method have very high potential to be able to extend the application fields to the optical excited states, the van-der Waals interactions between fragments in the molecular system and the high-temperature superconductor.

Acknowledgements

The authors would like to acknowledge the infra-structure support for the Joint Research Programs in Graduate School of Science and Technology, Meiji University, Japan.

Author details

Masami Kusunoki
Meiji University, Kawasaki, Japan

*Address all correspondence to: kusunoki@meiji.ac.jp

References

[1] Hohenberg P, Kohn W. Inhomogeneous electron gas. Physical Review. 1964;**136**(B):864-871. DOI: 10.1103/PhysRev.136.B864

[2] Kohn W, Sham LJ. Self-consistent equations including exchange and correlation effects. Physical Review. 1965;**140**(A):1133-1137. DOI: 10.1103/PhysRev.140.A1133

[3] Sham LJ. Exchange and correlation in density-functional theory. Physical Review B. 1985;**32**:3876-3882. DOI: 10.1103/PhysRevB.32.3876

[4] Parr RG, Yang W. Density-functional theory of atoms and molecules. In: Breslow R, Goodenough JB, Halpern J, Rowlinson JS, editors. International Series of Monographs on Chemistry. New York: Oxford Science; 1989. vol. 16, pp. 1-333. ISBN: 0-19-509276-7

[5] Dreizler RM, Gross EKU. Density Functional Theory: An Approach to the Quantum Many-Body Problem. Berlin: Springer; 1990. pp. 1-302. DOI: 10.1007/978-3-642-86105-5

[6] Kock W, Holthausen MC. A Chemist's Guide to Density Functional Theory. Weinheim: WILEY-VCH; 2000. pp. 1-294. ISBN: 3-527-29918-1

[7] Becke AD. Perspective: Fifty years of density-functional theory in chemical physics. Journal of Chemical Physics. 2014;**140**(18):18A301. DOI: 10.1063/1.4869598

[8] Sundararaman R, Goddard WA, Alias TA. Grand canonical electronic density functional theory: Algorithms and applications to electrochemistry. Journal of Chemical Physics. 2017; **146**:114104-114114. DOI: 10.1063/1.4978411

[9] Evangelista FA. Perspective: Multireference coupled cluster theories of dynamical electron correlation. Journal of Chemical Physics. 2018;**149**: 030901-030914. DOI: 10.1063/1.5039496

[10] Mardirossian N, Head-Gordon M. Thirty years of density functional theory in computational chemistry: An overview and extensive assessment of 200 density functionals. Molecular Physics. 2017;**115**:2315-2372. DOI: 10.1080/00268976.2017.1333644

[11] Kitakawa CK, Maruyama T, Oonari J, Mitsuta Y, Kawakami T, Okumura M, et al. Linear response functions of densities and spin densities for systematic Modeling of the QM/MM approach for mono- and poly-nuclear transition metal systems. Molecules. 2019;**24**:821-849. DOI: 10.3390/molecules24040821

[12] Zhou C, Hermes MR, Wu D, Bao JJ, Pandharkar R, King DS, et al. Electronic structure of strongly correlated systems: Recent developments in multiconfiguration pair density functional theory and multiconfiguration nonclassical-energy functional theory. Chemical Science. 2022;**13**:685-7706. DOI: 10.1039/d2sc01022d

[13] Kusunoki M. Heisenberg spin Hamiltonian derived from a multiple grand canonical spin density functional theory with a principal nonlocal exchange-correlation energy functional. Journal of Physical Society of Japan. 2022;**91**:014702-(1-28). DOI: 10.7566/JPSJ.91.014702

[14] Mulliken RS. Electronic population analysis on LCAO–MO molecular wave functions. I. Journal of Chemical Physics. 1955;**23**:1833-1840. DOI: 10.1063/1.1740588

[15] Lee C, Yang W, Parr RG. Development of the colic-Salvetti correlation-energy formula into a functional of the electron density. Physical Review B. 1988;**98**:785-789. DOI: 10.1103/PhysRevB.37.785

[16] Becke AD. A new mixing of Hartree-Fock and local density-functional theories. Journal of Chemical Physics. 1993;**98**: 1372-1378. DOI: 10.1063/1.464304

[17] Najafpour MM, Renger G, Holynska M, Moghaddam AN, Aro EM, Carpentier R, et al. Manganese compounds as water-oxidizing catalysts: From the natural water-oxidizing complex to Nanosized manganese oxide structures. Chemicl Reviews. 2016;**116**:2886-2936. DOI: 10.1021/acs.chemrev.5b00340

[18] Dexheimer SL, Klein MP. Detection of a paramagnetic intermediate in the S1 state of the photosynthetic oxygen-evolving complex. Journal of American Chemical Society. 1992;**114**:2821-2826. DOI: 10.1021/ja00034a010

[19] Yamauchi T, Mino H, Matsukawa T, Kawamori A, Ono T. Parallel polarization electron paramagnetic resonance studies of the S1-state manganese cluster in the photosynthetic oxygen-evolving system. Biochemistry. 1997;**36**:7520-7526. DOI: 10.1021/bi962791g

[20] Campbell KA, Gregor W, Pham DP, Peloquin JM, Debus RJ, Britt RD. The 23 and 17 kD extrinsic proteins of photosystem II modulate the magnetic properties of the S1-state manganese cluster. Biochemistry. 1998;**37**: 5039-5045. DOI: 10.1021/bi9800552

[21] Suga M, Akita F, Hirata K, Ueno G, Murakami H, Nakajima Y, et al. Native structure of photosystem II at 1.95A° resolution viewed by femtosecond X-ray pulses. Nature. 2015;**517**:99-103. DOI: 10.1038/nature13991

[22] Kusunoki M. S1-state Mn4Ca complex of photosystem II exists in equilibrium between the two most-stable isomeric substates: XRD and EXAFS evidences. Journal of Photochemistry and Photobiology B. 2011;**104**:100-1010. DOI: 10.1016/j.jphotobiol.2011.03.002

[23] Bardeen J, Cooper LN, Schrieffer JR. Theory of superconductivity. Physical Review. 1957;**108**:1175-1204. DOI: 10.1103/PhysRev.108.1175

[24] Schrieffer JR. Theory of Superconductivity. New York: WA Benjamin, Inc; 1964. pp. 1-282 ISBN: 9780429975332

[25] Fröhlich H. Theory of the superconducting state. I. the ground zero of temperature. Physical Review. 1950; **79**:845-846. DOI: 10.1103/PhysRev. 79.845

[26] Bogoliubov NN. A new method in the theory of superconductivity. Soviet Physics JETP. 1958;**34**:41-46

[27] Drozdov AP, Eremets MI, Troyan IA, Ksenofontov V, Shylin SI. Conventional superconductivity at 203 kelvin at high pressures in the sulfur hydride system. Nature. 2015;**525**:73. DOI: 10.1038/nature14964

Section 2

DFT Application

Chapter 4

Monitoring Organic Synthesis *via* Density Functional Theory

Nurdiana Nordin

Abstract

A preliminary molecular structure for a system, which may or may not be known, is the first step in a typical investigation using ab initio techniques. A stable system is generated by a geometry search using an energy minimization method (usually a local minimum or transition state). Subsequently, it is easy to obtain any energetic properties (such as atomization energies, formation temperatures, binding energies) or expectation values or quantifiable quantities from the wave function of the molecular system and its fragments. The stability of such a system can be determined by considering the second derivative of the energy with respect to the spatial coordinates (also known as the Hessian matrix). It could be a goal to find out how the system interacts with other systems and eventually to decipher the synthesis pathways. Therefore, this chapter presents a recent application of approaches based on density functional theory (DFT) to study chemical processes at the catalytic sites of enzymes. The focus is on the interaction of small organic molecules with the ability to inhibit a catalytic cysteine of the malaria parasite, in the area of drug design.

Keywords: density functional theory, ab initio, organic reaction, local optimization, transition states, potential energy surface, enzymatic catalysis

1. Introduction

Two quantum-chemical theories have been established since Lewis first proposed the idea of a chemical bond at the beginning of the twentieth century [1]: the valence bond (VB) theory and the molecular orbital (MO) theory, both of which are based on Schrödinger's equation [2]. The density functional theory (DFT), which states that the ground state energy of a non-degenerate N-electron system is a special functional of the density (r), was developed in the 1960s of the previous centuries on the basis of the Hohenberg and Kohn theorems.

$$E[\rho(r)] = \int \rho(r)v(r)\,dr + F[\rho(r)] \tag{1}$$

The "external one-electron potential" or the electron-nucleus Coulomb interaction is represented by v(r), and F[(r)] is the Hohenberg-Kohn universal functional obtained by adding the kinetic energy functional T[(r)] and the energy functional of

IntechOpen

the electron-electron interaction $V_{ee}[(r)]$. The rigorous theoretical basis of DFT is this theorem. If the number of electrons is kept constant within the DFT framework, the electron density can be represented as a functional derivative of the energy with respect to the external potential:

$$\boldsymbol{\rho}(\boldsymbol{r}) = \left(\frac{\boldsymbol{\delta} E}{\boldsymbol{\delta}\, v(\boldsymbol{r})}\right)_{\mathrm{N}} \tag{2}$$

Density functional theory (DFT) calculations therefore require the construction of an expression for the electron density. Similar to the quantum chemical theory based on the Schrödinger equation, it is computationally impossible to resolve the electron density functional $\rho(r)$ for a complicated system. The definition of the individual terms in the functional $F[\rho]$ is the crux of the mathematical problem. The Hartree-Fock equation served as a rough analogy to introduce the Kohn-Sham formalism [3]. In recent years, numerous empirical DFT functions have been developed, including B3LYP [4, 5], MPWB1K [6], and more recently M06 and related functions [7], which provide precise energies and allow the study of organic reactions with a computational cost comparable to MO calculations.

A thorough quantum chemical study of molecular electron density in terms of nonbinding and binding molecular areas was made possible by the invention of topological analysis of the electron localization function in the late twentieth century [8]. This investigation made it possible to build a molecular model that was connected to the Lewis binding pattern. The molecular mechanism of the majority of organic reactions has been identified through topological electron localization function analysis of binding alterations throughout a reaction pathway. The construction of a reactivity model in which these bonds are produced by the C-to-C coupling of two pseudo radical centers [9] formed along the reaction pathway [10] has been made possible by several studies of organic reactions in which C-C bonds are formed. This pattern is interestingly present in C-C double bond reactions that are nonpolar, polar, and ionic. High activation energies in nonpolar reactions are a result of the energy needed to break C-C double bonds in order to produce pseudo radical structures. It is noteworthy that when the reaction's polarity rises, these high activation energies diminish. Domingo has shown that the global electron density transfer that takes place in polar processes favors the variations in electron density necessary for the creation of C-C single bonds.

An essential and typical first step in most quantum chemical studies is structural optimization. It is a crucial element of any computational chemistry study that deals with the structure and reactivity of a molecule. There are numerous methods for structural optimization. These techniques are used to find transition structures (TS), find minimum energy paths (MEP) that correlate with reaction pathways, and optimize equilibrium geometries. In this chapter, a concise protocol is presented to understand how to easily obtain energetic properties (such as atomization energies, formation temperatures, binding energies) by considering the first and second derivatives of the energy with respect to the Hessian matrix.

1.1 Local optimization

The last 50 years have seen rapid progress in optimization techniques for ab initio molecular orbital simulations. The introduction of energy gradient techniques, the development of algorithms, and the increase in processing capacity have played an important role [11–13]. Optimization algorithms based solely on energy are orders of magnitude slower than analytical gradient-based optimization strategies. Searching for transition structures is now feasible, and optimization of equilibrium geometries has become routine, even for quite large systems.

1.1.1 Potential energy surface

The relative positions of the individual atoms within a molecule define its molecular structure. The molecule has a clear energy for a particular location and electronic state. The potential energy surface (PES) describes this energy, which varies depending on the electronic state and atomic coordinates. The potential energy surface of two geometric variables is roughly shown in **Figure 1**. The Born-Oppenheimer approximation [15], in which the motions of nuclei and electrons are studied independently, leads to the ideas of potential energy surfaces. Nuclei move slower than electrons because they are much heavier. This makes it possible to separate the nuclear motions from the electronic motions.

The minima, maxima, and saddle points—the stationary points that form the surface of the potential energy—define it [16]. Each point, which can be identified from the first and second derivatives of the molecule, indicates a different state of the molecule. The first energy derivatives of each atom with respect to its coordinates combine to give a vector called a gradient. The combined second derivatives are used to form a matrix called a Hessian. A vanishing gradient is a property of any stationary point. Also, the Hessian matrix is positively defined at a minimum (all eigenvalues of the Hessian matrix are positive) and the Hessian matrix has only one negative eigenvalue at a first-order saddle point. In chemistry, minima indicate stable structures, while first-order saddle points can be associated with transition states (TS). For structural optimization, higher-order saddle points on the potential energy surface are not relevant (but they are relevant for electronic structure calculations). When using mass-weighted coordinates, the minimum energy path (MEP) or intrinsic reaction coordinate (IRC) consists of the steepest descent reaction paths (SDP) from the

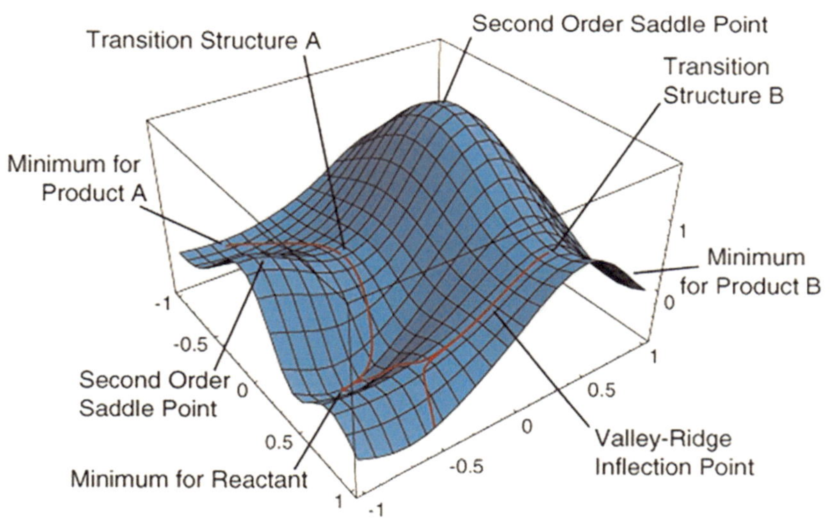

Figure 1.
The minima, saddle points, and inflection points are displayed as an intriguing feature of the PES model [14].

transition state to the minima on either side of the saddle point. Taylor expansion allows to represent the potential energy surface, $E(\boldsymbol{x})$, as an infinite sum in neighborhood of a point $\boldsymbol{x}_O$, using the step $\boldsymbol{x}$ and gradient $\boldsymbol{g}^T$ vectors, and Hessian matrix, $\boldsymbol{H}$,

$$E(\boldsymbol{x}) = E(\boldsymbol{x}_0) + \boldsymbol{g}^T \boldsymbol{x} + 1/2\, \boldsymbol{x}^T \boldsymbol{H}_x + \ldots \quad (3)$$

Most optimization methods are constructed based on this representation. Since two coordinate systems connected by a linear transformation are equivalent in all gradient optimization methods, the significance of the coordinates for the optimization process may not be immediately apparent. This is because the gradient vector and the Hessian matrix must be properly transformed from one system to the other. On this basis, it has been argued that the Cartesian coordinates are comparable to the valence-type internal coordinates [17–19] or even better [20]. This claim [17–19] ignores the potential importance of cubic and higher valence couplings. However, it has been shown that optimization in Cartesian coordinates can be as effective as in Z-matrix coordinates [20] or natural internal coordinates for medium-sized systems with a trustworthy initial Hessian matrix and suitable initial geometries [21]. However, optimization in Cartesian coordinates is incredibly wasteful, especially if the system is large enough, since it lacks initial curvature information (i.e., without a Hessian matrix) [20, 21]. Even if the initial geometries are suboptimal for optimization, Cartesian coordinates are less effective than internal coordinates, even if the exact Hessian matrix is always known [22]. A good and affordable initial Hessian matrix for ab initio calculations is usually a molecular mechanics force constant matrix.

1.2 Monitoring transition states

Numerous methods have been proposed to find suitable starting geometries because the calculations of the transition structures are so sensitive to the starting geometry. One very helpful method is to start with reactant and product structures, which are easier to obtain than transition structures. The simplest way to determine the shape of a transition structure is to assume that each atom is exactly halfway between its initial and final positions. The term "linear synchronous transit" refers to this nearly linear motion (LST). Although this is a good first approximation, it is not perfect. Consider the motion of an atom whose bond angle varies relative to the other molecules in the system. The bond length in the middle of the line connecting its starting and ending points will be shorter than expected and therefore have a larger energy (perhaps much larger). The quadratic synchronous transit (QST) method is the logical development of this technique. These techniques assume that a parabola connecting the geometries of reactant and product is formed by the positions of the atoms in the transition structure. Even if it is only a very small improvement, QST is usually better than LST. A weighting factor can often be entered by the user (e.g., to specify a structure containing 70% products and 30% reactants). This allows the application of the Hammond postulate, which states that the transition structure resembles the reactants in an exothermic reaction or the products in an endothermic reaction [23]. These methods have their limitations, but have proven to be quite useful for simple reactions. The drawback is that each of these methods, however good, is based on the premise that the reaction proceeds in a single step with coordinated motion of all atoms [24]. For multistep reactions, each of these methods can be used on its own. For a reaction with a single transition structure but uncoordinated motion (e.g., breaking one bond and building another), the use of hand-drawn initial geometries or eigenvalue tracking may be preferable.

1.2.1 Reaction coordinate

The maximum on the reaction pathway is by definition a transition structure. There is a single, well-defined reaction pathway, namely the intrinsic or low energy (IRC) pathway. Numerous groups have developed methods to derive the IRC pathway from quasi-Newton optimization. These methods are based on the observation that quasi-Newton methods oscillate around the IRC path from one iteration to the next.

By following the reaction path from the equilibrium geometry to the transition structure, one can also obtain a transition structure. The user chooses which mode of oscillation to trigger a reaction given sufficient kinetic energy, which is why this method is also known as eigenvalue tracking [25]. This is not the most effective approach to obtaining an IRC, nor is it the fastest or most reliable way to find a transition structure. The advantage is that there are no assumptions about the transition structure or the coordinated motions of the atoms.

Using a pseudo-reaction coordinate is another method. This method can be a lot of work for the user and takes more time than most other methods. However, it has the advantage of being extremely reliable, so it will work even when all other methods have failed. By first selecting a geometric parameter that is directly related to the reaction, a pseudo-reaction coordinate is calculated (e.g., the bond length for a bond that is formed or broken). Then, a series of calculations is performed while all other geometric parameters are optimized and this parameter is fixed at various values ranging from those in the reactants to those in the products.

The result is an approximation of the reaction coordinate rather than the actual reaction coordinate, which is close to the actual reaction coordinate only for equilibrium geometries and transition structures. Normally, the geometry for a quasi-Newton optimization starts with the calculation from this set with the highest energy. Quasi-Newton optimization may still fail in some rare cases where extremely flat potential surfaces are present. In this case, the transition structure can be estimated with arbitrary accuracy (within the theoretical model) by determining the maximum energy by adjusting the selected geometric parameter in smaller and smaller steps.

1.2.1.1 Solvent effects

The choice of solvent can influence the reaction rate. Interactions with the solvent often change the geometry of the transition structure only slightly, but have a large effect on the energy of the structure. If the solvent effects are taken into account in the calculation, all transition structure discovery methods can be applied. The transition structure identification method is not affected by the presence of solvent interactions.

1.2.2 Evaluating the Hessian

Partitioned-rational function optimization (P-RFO), a useful technique, is based on partitioning the eigenvalues of the Hessian into modes with negative curvatures along which the energy is maximized and all other modes along which the energy is minimized. To guarantee convergence to the nearest transition states, P-RFO requires an initial Hessian with a single negative eigenvalue along the reaction coordinate. If all eigenvalues of the hessian are positive, the P-RFO search may not find the desired transition states because the smallest positive eigenvalue is selected as the mode for energy maximization, which may not match the reaction coordinate. The same problem occurs when the Hessian has numerous negative eigenvalues of comparable magnitude. In this case, the most negative eigenvalue is tracked upward, even though it may not be the reaction coordinate. Even if the transition states estimate is accurate, a P-RFO calculation may fail if the Hessian is not calculated with high accuracy, since inaccuracies in the sign of the eigenvalues create the same ambiguity in determining which mode to follow upward [26]. Monitoring a reaction rate using the activation energy in the Arrhenius equation is the simplest technique for determining a reaction rate. Experimental results or a simple theoretical approach, such as the kinetic theory of gasses, can be used to determine the pre-exponential factor. The activation energy can be roughly calculated by subtracting the energy of the reactant and transition structures. The addition of the zero-point vibrational energy leads to an easily determined additional correction to these energies.

The simple use of activation energy implies that the intrinsic reaction coordinate is the only direction in which a reaction can proceed. It would be more accurate to consider the possibility that reactions can also occur that pass through a geometry that closely approximates the transition structure. The variable transition state calculations take this into account. The vibrational frequencies of the transition structure, the full reaction coordinate, or the total potential energy surface may need to be used in these calculations. Tunneling at the reaction barrier may also be considered in these calculations. These calculations can provide accurate answers, but they are particularly susceptible to minute parameters such as the choice of a mass-weighted coordinate system for the geometry.

Dynamic analyses can be used to study how the direction and velocity of the entering reactants affect the reaction rate. These studies often begin with ab initio calculated potential energy surface. Although obtaining the potential energy surface was not an easy task in itself, the effort required to study a reaction using these methods can be significantly greater.

2. Structural model based on computational method

One of the conventional methods for modeling matter based on quantum mechanics, including atoms, molecules, solids, nuclei, and both quantum and classical fluids, is density functional theory (DFT). The first category includes techniques often referred to as ab initio techniques, such as Hartree-Fock (HF), Moller-Plesset perturbation theory (MP), configuration interaction (CI), and linked-cluster methods (CC) [27]. Given that the wave function for a system of N electrons depends on 3N spatial variables, it should be noted that as the number of electrons increases, so does the complexity of the wave function. DFT approaches [28], on the other hand, use functionals of the electron density which are a function of only 3N spatial variables and independent of the system size.

The first H-K theorem shows how the Hamilton operator, and consequently all the properties of the system, is uniquely determined by the electron density, while the second theorem asserts that the functional related to the ground-state energy of the system yields the lowest energy if and only if the input density is the true ground-state density (i.e., nothing other than the variational principle). Currently, the correctness and effectiveness of DFT-based approaches depend on the basis chosen for the expansion of the Kohn-Sham orbitals, but in particular on the caliber of the exchange correlation (XC) functional used [3].

Local density approximation (LDA) was the first method originally proposed for modeling XC functionals [29]. A simple generalization of the local spin density approximation (LDA) to include electron spin in the functional was proposed by Slater [30]. The well-known SVWN functional, whose exchange component was created by Slater [31] and whose correlation component was created by Vosko, Wilk, and Nusair [31] is indeed one of the early LSDA-XC functionals. However, experience has shown that LSDA leads to inflated binding energies and underestimation of barrier heights, although it usually provides the bond lengths of molecules and solids with an amazing accuracy (about 2%). Thus, the average accuracy of LSDA is insufficient for most applications in monitoring organic synthesis in chemistry.

Compared to LSDA functions, GGA (Generalized Gradient Approximation) functions have been shown to provide better predictions for total energies, atomization energies, and structural properties. The most commonly used XC functions of this type are the Perdew 86 (P86), Becke (B86, B88), Perdew-Wang 91 (PW91), Laming-Termath-Handy (CAM), Perdew-Burke-Ernzerhof (PBE), revised Perdew-Burke-Ernzerhof (RPBE), Perdew-Burke-Ernzerhof revised for solids (PBEsol), Becke exchange and Lee-Yang-Parr correlation (BLYP), and Armiento-Mattsson 2005 (AM05). However, they still provide too low barrier heights and usually fail in describing van der Waals interactions.

An alternative strategy has been developed, called the SCC-DFTB (self-consistent-charge density functional tight binding, later abbreviated as DFTB) method. The DFTB approach, which is an alternative to the traditional semi-empirical methods in quantum chemistry, including the well-known MNDO, AM1, and PM3 schemes derived from RF theory, is based on a second-order expansion of the DFT total energy expression. Since its parameterization technique is based entirely on DFT calculations and does not require fitting empirical data, DFTB is not, strictly speaking, a semi-empirical method.

3. Case study: monitoring enzymatic inhibition

Typically, a two-step mechanism is used to describe how the enzyme and substrate function. First, favorable placement of the molecule in the binding pocket ensures the formation of a non-covalent enzyme-inhibitor complex. When the thiolate group of the catalytic CYS attacks one of the carbon atoms of the ring, ring opening occurs (**Figure 2**), leading to irreversible alkylation of this amino acid.

The effect of the environment at the carbon atoms in the rings on the kinetics and thermodynamics was evaluated by Helten and colleagues using quantum chemical simulation [32, 33]. According to these estimates, acidic media have a greater effect on the opening of the aziridine ring than for epoxide. They discovered that the protonation of the nitrogen center of aziridine-based inhibitors occurs at the beginning of the reaction course, more precisely before the transition state for the ring opening. It was also discovered, and this is important, that aziridine is inactive without the prior N-protonation step. Therefore, it is suggested that protonation significantly accelerates the reaction rate by stabilizing both the transition state and the ring opening product. Hence, substituents that deprive the nitrogen atom of electrons should lead to lower reaction barriers and, consequently, a greater inhibition effect [32]. Since the products of this reaction are highly stabilized, only the thermodynamics of this reaction are favored by O-protonation, while the kinetics remain unchanged, in contrast to the behavior of aziridines [32]. This is due to the fact that, in the case of epoxide-based

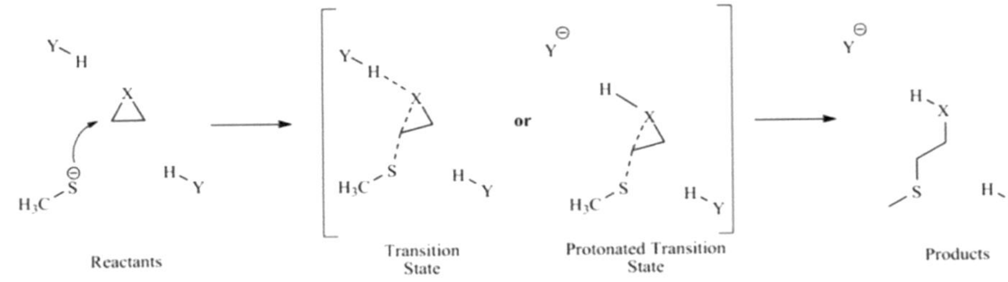

Figure 2.
Schematic representation of the ring opening mechanism of the alkylation of a methyl thiolate by an N-substituted aziridine ring. X = O, N-R. Adapted from [27].

inhibitors, the protonation of the oxygen center occurs only after the appearance of the transition state. The attacking cysteine was represented by a methyl thiolate (H_3C-S-), while the inhibitors were represented by tripartite ring systems $(H_2C)_2X$ with X = O, N-R. The authors characterized this simplified model as typical for the inhibition mechanism of the enzyme. The heteroatom of the three-membered rings and the methyl thiolate were placed near solvent molecules with increasing proton donor capacity to capture the effects of decreasing pH on the reaction profile. Water molecules were used to model settings with weak proton donor capabilities (pKa = 15.74). In contrast, NH_4^+ (pKa 9.3) and HCO_2H (pKa 3.8) molecules were used to mimic environments with stronger proton donor capabilities [33]. While the energies were derived using B3LYP [21] single point calculations, the geometry optimizations of the relevant stationary points, verified by frequency calculations, were calculated using the BLYP [34, 35] functional. A TZVP basis set and both functionals were merged [36]. The accuracy of the theoretical approach was evaluated and the researchers found that BLYP significantly underestimated the barrier heights, while the response profile derived with B3LYP showed excellent agreement with CCSD (T) data [32].

A schematic representation of the mechanism of action at the catalytic site is formed by a cysteine and a histidine, whose side chains form a thiolate/imidazolium ion pair, and an asparagine, which plays a crucial role in the appropriate alignment of the ion pair [37], is outlined in **Figure 3**. One of the key events in the catalytic hydrolysis of hemoglobin is the nucleophilic attack of the thiolate anion on the corresponding electron-deficient carbonyl group of the substrate, resulting in a negatively charged tetrahedral intermediate stabilized by the "oxyanion hole" formed by the side chains of GLN36 and TRP206 in FP2 and by GLN38 and TRP208 in FP3 [38]. The kinetics, thermodynamics, and regioselectivity of the ring-opening reaction of epoxide- and aziridine-based compounds can be evaluated by using standard quantum mechanics (QM) calculations [32, 33].

The enzyme, the warhead of the inhibitor, and water are the components of the model system used for the quantum mechanics/molecular mechanics (QM/MM) calculations. The part subjected to quantum mechanics corresponded to the zwitterionic side chains of the catalytic residues CYS and HIS and the electrophilic warhead of the inhibitors. The calculations in the QM domain both with and without the mediating water molecule can identify if water could mediate proton transfer. One-point calculations at the B3LYP/TZVP level of theory are used to more accurately determine the activation and reaction energies, while QM calculations can be performed during geometry optimization and potential energy surface scanning. The MM part can be

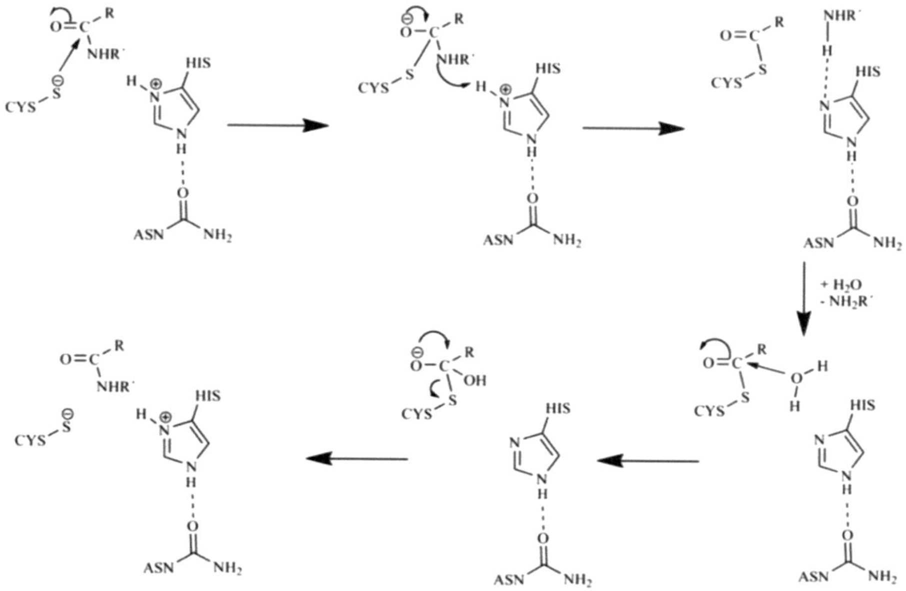

Figure 3.
Schematic reaction mechanism of the cysteine protease-mediated cleavage of a peptide bond.

implemented using the CHARMM force field [39]. The inhibition reaction, which included the ring opening of the inhibitor and the formation of the S(CYS)-C bond, can be captured by the calculated potential energy surface. The results rule out a direct proton shift from the HIS to the organic ligand and show that even a single water molecule is sufficient to create a highly effective relay system that allows protons to move from the HIS to the inhibitor. This result strongly suggests that the effectiveness of the inhibitors can be reduced by substituents that can prevent proton transfer.

The speed-up of the calculations can be even greater when a parameterized quantum mechanics technique such as the density functional tight binding (DFTB) approach is considered or when it is combined with molecular mechanics. This makes QM/MM MD simulations in the ns range easily feasible, which is an essential requirement for reliable determination of free energies for biomolecular reactions.

4. Conclusions

The purpose of this chapter is to present a brief protocol of conceptual DFT reactivity in enzymatic reactions. The calculations can be used to easily determine energetic properties (such as atomization energies, formation temperatures, binding energies) by considering the first and second derivatives of the energy with respect to the Hessian matrix. Chemical intuition comes into play during the procedure, such as understanding the shape of transition states in a system related to the system under study. It is likely that the transition states can be found if the conjecture is strong enough to indicate that it lies in the quadratic basin of transition states, or in other words, has a negative Hessian eigenvalue. The main advantage of this homolog-based method is that it is computationally free and effective when dealing with simple or very closely related systems with known transition states. On the other hand, with the help of realistic models and DFT techniques, enzymatic catalysis is becoming more understandable at the electronic level. This is particularly helpful for the development of irreversible enzyme inhibitors that can covalently bind the catalytic amino acids of the enzyme.

Acknowledgements

N.N acknowledges funding from the Ministry of Higher Education Malaysia for FRGS grant, KPT (FP118-2020)/1/2020/STG04/UM/01/1, and RU grant (ST018-2021).

Conflict of interest

The authors declare no conflict of interest.

Author details

Nurdiana Nordin
Faculty of Science, Department of Chemistry, University of Malaya, Kuala Lumpur, Malaysia

*Address all correspondence to: ndiana13@um.edu.my

References

[1] Lewis GN. The atom and the molecule. Journal of the American Chemical Society. 1916;**38**(4):762-785

[2] Schrödinger E. An undulatory theory of the mechanics of atoms and molecules. Physical Review. 1926;**28**(6): 1049

[3] Kohn W, Sham LJ. Self-consistent equations including exchange and correlation effects. Physical Review. 1965;**140**(4A):A1133

[4] Becke AD. Density-functional thermochemistry. III the role of exact exchange. The Journal of Chemical Physics. 1993;**98**:5648-5652

[5] Mei Y, Chen Z, Yang W. Exact second-order corrections and accurate quasiparticle energy calculations in density functional theory. Journal of Physical Chemistry Letters. 2021;**12**(30): 7236-7244

[6] Zhao Y, Truhlar DG. Hybrid meta density functional theory methods for thermochemistry, thermochemical kinetics, and noncovalent interactions: The MPW1B95 and MPWB1K models and comparative assessments for hydrogen bonding and van der Waals interactions. The Journal of Physical Chemistry. A. 2004;**108**(33):6908-6918

[7] Zhao Y, Truhlar DG. The M06 suite of density functionals for main group thermochemistry, thermochemical kinetics, noncovalent interactions, excited states, and transition elements: Two new functionals and systematic testing of four M06-class functionals and 12 other functionals. Theoretical Chemistry Accounts. 2008;**120**(1):215-241

[8] Becke AD, Edgecombe KE. A simple measure of electron localization in atomic and molecular systems. The Journal of Chemical Physics. 1990;**92**(9): 5397-5403

[9] Domingo LR, Chamorro E, Perez P. Understanding the high reactivity of the azomethine ylides in [3 + 2] cycloaddition reactions. Letters in Organic Chemistry. 2010;**7**(6):432-439

[10] Domingo LR. A new C-C bond formation model based on the quantum chemical topology of electron density. RSC Advances. 2014;**4**(61): 32415-32428

[11] Pulay P. ab initio calculation of force constants and equilibrium geometries in polyatomic molecules: I. theory. Molecular Physics. 1969;**17**(2):197-204

[12] Bishop DM, Randič M. ab initio calculation of harmonic force constants. The Journal of Chemical Physics. 1966; **44**(6):2480-2487

[13] Gerratt J, Mills IM. Force constants and dipole-moment derivatives of molecules from perturbed Hartree-Fock calculations I. The Journal of Chemical Physics. 1968;**49**(4):1719-1729

[14] Schlegel HB. Geometry optimization. Wiley Interdisciplinary Reviews: Computational Molecular Science. 2011;**1**(5):790-809

[15] Born M, Heisenberg W. Zur quantentheorie der molekeln. In: Original Scientific Papers Wissenschaftliche Originalarbeiten. Switzerland AG: Springer Nature; 1985. pp. 216-246

[16] Helgaker T. Optimization of minima and saddle points. In: Lecture Notes in Quantum Chemistry. Switzerland AG: Springer Nature; 1992. pp. 295-324

[17] Head JD, Weiner B, Zerner MC. A survey of optimization procedures for stable structures and transition states. International Journal of Quantum Chemistry. 1988;**33**(3):177-186

[18] Head JD, Zerner MC. A Broyden—Fletcher—Goldfarb—Shanno optimization procedure for molecular geometries. Chemical Physics Letters. 1985;**122**(3):264-270

[19] Head JD. Partial optimization of large molecules and clusters. Journal of Computational Chemistry. 1990;**11**(1): 67-75

[20] Baker J, Hehre WJ. Geometry optimization in Cartesian coordinates: The end of the Z-matrix? Journal of Computational Chemistry. 1991;**12**(5): 606-610

[21] Baker J. Techniques for geometry optimization: A comparison of Cartesian and natural internal coordinates. Journal of Computational Chemistry. 1993; **14**(9):1085-1100

[22] Baker J, Chan F. The location of transition states: A comparison of Cartesian, Z-matrix, and natural internal coordinates. Journal of Computational Chemistry. 1996;**17**(7):888-904

[23] Young D. Finding transition structures. In: Computational Chemistry: A Practical Guide for Applying Techniques to Real World Problems. New York, NY: John Wiley & Sons, Inc; 2002. pp. 145-158

[24] Jensen F. Transition structure modeling by intersecting potential energy surfaces. Journal of Computational Chemistry. 1994;**15**(11): 1199-1216

[25] Hehre WJ. Practical Strategies for Electronic Structure Calculation. Irvine: Wavefunction, Inc.; 1995. pp. 102-134

[26] Keil FJ. Multiscale modelling in computational heterogeneous catalysis. Topics in Current Chemistry. 2012;**307**: 69-107

[27] Burkert U, Allinger NL. Pitfalls in the use of the torsion angle driving method for the calculation of conformational interconversions. Journal of Computational Chemistry. 1982;**3**(1):40-46

[28] Jensen F. Introduction to Computational Chemistry. Hoboken, New Jersey, USA: John Wiley & Sons, Inc.; 2017

[29] Dirac PAM. Note on exchange phenomena in the Thomas atom. Mathematical Proceedings of the Cambridge Philosophical Society. 1930; **26**(3):376-385

[30] Slater JC. The Self-Consistent Field for Molecules and Solids: Quantum Theory of Molecules and Solids. McGraw-Hill Book Company, Australia Holdco Pty. Ltd; 1974

[31] Vosko SH, Wilk L, Nusair M. Accurate spin-dependent electron liquid correlation energies for local spin density calculations: A critical analysis. Canadian Journal of Physics. 1980;**58**(8):1200-1211

[32] Helten H, Schirmeister T, Engels B. Theoretical studies about the influence of different ring substituents on the nucleophilic ring opening of three-membered heterocycles and possible implications for the mechanisms of cysteine protease inhibitors. The Journal of Organic Chemistry. 2005;**70**(1): 233-237

[33] Helten H, Schirmeister T, Engels B. Model calculations about the influence of protic environments on the alkylation step of epoxide, aziridine, and thiirane based cysteine protease inhibitors. The

Journal of Physical Chemistry. A. 2004; **108**(38):7691-7701

[34] Xiao H, Tahir-Kheli J, Goddard WA III. Accurate band gaps for semiconductors from density functional theory. Journal of Physical Chemistry Letters. 2011;**2**(3):212-217

[35] Perdew JP, Yue W. Accurate and simple density functional for the electronic exchange energy: Generalized gradient approximation. Physical Review B. 1986;**33**(12):8800

[36] Schäfer A, Huber C, Ahlrichs R. Fully optimized contracted Gaussian basis sets of triple zeta valence quality for atoms Li to Kr. The Journal of Chemical Physics. 1994;**100**(8): 5829-5835

[37] Leung D, Abbenante G, Fairlie DP. Protease inhibitors: Current status and future prospects. Journal of Medicinal Chemistry. 2000;**43**(3):305-341

[38] Teixeira C, Gomes JRB, Gomes P. Falcipains, plasmodium falciparum cysteine proteases as key drug targets against malaria. Current Medicinal Chemistry. 2011;**18**(10):1555-1572

[39] MacKerell AD Jr et al. All-atom empirical potential for molecular modeling and dynamics studies of proteins. The Journal of Physical Chemistry. B. 1998;**102**(18):3586-3616

Chapter 5

Impact of Crystal Parameters in XRD and DFT Measurements

Subramanian Usha and Charles Kanakam Christopher

Abstract

The Zwitterionic property of aminoacids give molecular crystal formation through homodesmotic reaction with smaller organic molecules which can undergo hydrogen bonding interactions. Alpha hydroxyl phenyl acetic acid known as mandelic acid (MA) was added with essential amino acid, L-phenylalanine (LPA) resulted in the formation of molecular crystal with $P2_1$ space group otho rhombic crystal containing four units (namely one MA, two LPA and one water) bis-L –phenyl alanine mandelate (BLPAMA) by slow evaporation method. The single crystal obtained was subjected to characterisation studies. Recrystallised BLPAMA using methanol, subjected to slow evaporation method resulted in the formation of non centerosymmetric C2 point group monoclinic single crystal of R-phenylalanine-S-mandelate (RPASMA) confirmed with XRD study. The theoretical DFT study of RPASMA using Gaussian 09 software to study the non-covalent interactions with MO6,6-31++G(d,p) showed encouraging results for the formation of low energy gap, highly reactive RPASMA. The H-bonding in the crystal confirmed by DFT study showed the existence of three units – MA, H and LPA in the crystal. Compared the experimental and theoretical crystal parameters of the reactants (MA, LPA) and product (RPASMA) for the thermo chemical properties, intermolecular hydrogen bonding existing between MA and LPA stabilises the structure of the formed RPASMA crystal resulting in the small difference in energy gap observed from HOMO-LUMO studies indicate the highly reactive character of RPASMA.

Keywords: crystal parameters, theoretical study, intermolecular hydrogen bonding, low energy gap, thermo chemical properties

1. Introduction

MA has a long history of use in the medical community as an antibacterial, particularly in the treatment of urinary tract infections. It has also been used as an oral antibiotic and as a component of chemical face peels analogous to other alpha hydroxy acids. LPA is an essential aromatic amino acid in humans (provided by food), LPA plays a key role in the biosynthesis of other amino acids and is important in the structure and function of many proteins and enzymes. LPA is converted to tyrosine, used in the biosynthesis of dopamine and norepinephrine neurotransmitters.

The LPA is incorporated into proteins, while the D-form acts as a painkiller. Absorption of ultraviolet radiation by Phenylalanine (PA) is used to quantify protein amounts.

LPA is the L-enantiomer of PA. It has a role as a nutraceutical, a micronutrient, an Escherichia coli metabolite, a Saccharomyces cerevisiae metabolite, a plant metabolite, an algal metabolite, a mouse metabolite, a human xenobiotic metabolite and an EC 3.1.3.1 (alkaline phosphatase) inhibitor. It is an erythrose 4-phosphate/phosphoenolpyruvate family amino acid, a proteinogenic amino acid, a phenylalanine and a L-alpha-amino acid. LPA is a conjugate base of a L-phenylalaninium, conjugate acid of a L-phenylalaninate, an enantiomer of a D-phenylalanine and a tautomer of a L-phenylalanine zwitterion.

Homodesmic reaction of MA and LPA results in the two/four components molecular single crystals due to the zwitter ionic form of phenyl alanine. The formation of non linear crystals having low energy gap highly reactive phenylalanine mandelates are stabilised due to the intermolecular H-bonding and intramolecular hydrogen bonding along with the van der Waals forces of attraction between the molecules in the crystal formation [1, 2]. The hydrogen bonding interactions induces the polarisability character and dipole moments of the compounds are resulting in the application of the compounds in different areas like opto electronics property, biological activity, antioxidant activity etc., [3–5]. The experimental solid state measurements are compared with the theoretical gaseous state measurement for the crystal parameters, FTIR, FT Raman, thermochemical properties to confirm the applications of the product and the reactants. Nonlinear nature of the compounds comply with the expected vibrational frequencies calculated using the formula (3N-6) with the theoretical vibrational changes contribution to Total Energy Distribution (TED) [6]. The charge transfer mechanism in the molecule and the presence of high antioxidant power and antiradical power of RPASMA is confirmed by Homo-Lumo small energy gap, electrochemical Cyclic Voltametric analysis [7, 8].

2. Materials and methods

The compounds MA and LPA were taken to obtain the two component molecular crystal RPASMA by two steps mentioned below.

2.1 Step – 1: Synthesis of BLPAMA

Alfa Aesar DL- Mandelic acid(99%) and Nice chemicals *L*-phenylalanine in the molar ratio 1:2 respectively were taken in a beaker, dissolved in water, stirred well in a magnetic stirrer and obtained the molecular crystal BLPAMA by slow evaporation at room temperature. Confirmed the crystal structure by characterisation studies.

2.2 Step – 2: Synthesis of RPASMA

BLPAMA crystals were dissolved in 1:10 mmol in AR grade methanol, filtered, a clear pale yellow solution was obtained. Filterate was allowed to slow evaporation at

DOI: http://dx.doi.org/10.5772/intechopen.112291

RT, resulted in yellow coloured molecular crystals and confirmed the structure by characterisation studies.

Non-centerosymmetric crystal information of RPASMA given in CCDC number 1452128.

3. Computational analysis

Gaussian 09 with MO6 DFT 6-31++G(d,p) basis set shows the existence of hydrogen bonding charge transfer mechanism in the molecular crystal formation resulting in the formation of non-centerosymmetric structure having C2 space group [9]. Experimental parameters in the solid crystalline state XRD is compared with the gaseous phase density functional measurement using Gaussian 09 with MO6 DFT 6-31 ++G(d,p) basis sets are compared and discussed in detail [10]. The optimised DFT structure of the compounds are shown in the **Figures 1–3**.

The interactions in the title compounds, types of electrons, protons, vibrations are compared and discussed in detail [11]. Comparison of thermo chemical properties of the compounds show the energy involved during the crystal formation [12]. The electron transfer from the LPA molecule Highest occupied molecular orbital (HOMO) interaction with the lowest unoccupied orbital (LUMO) of MA molecule is confirmed from the theoretical DFT study supports the experimental study of the compounds [13].

As the size of the molecule increases the charge and dipole moment increase and hence polarisability increases as shown in **Table 1**.

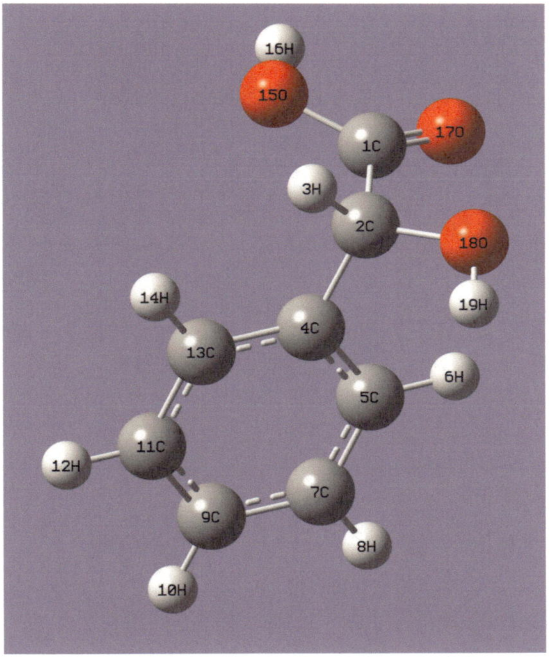

Figure 1.
Labelled MA.

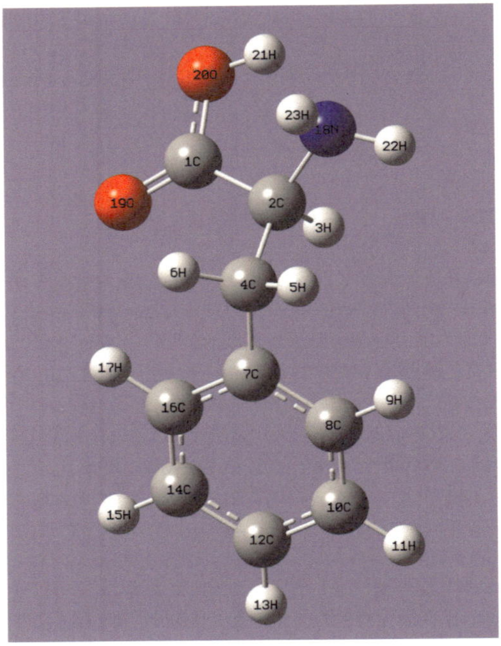

Figure 2.
Labelled LPA.

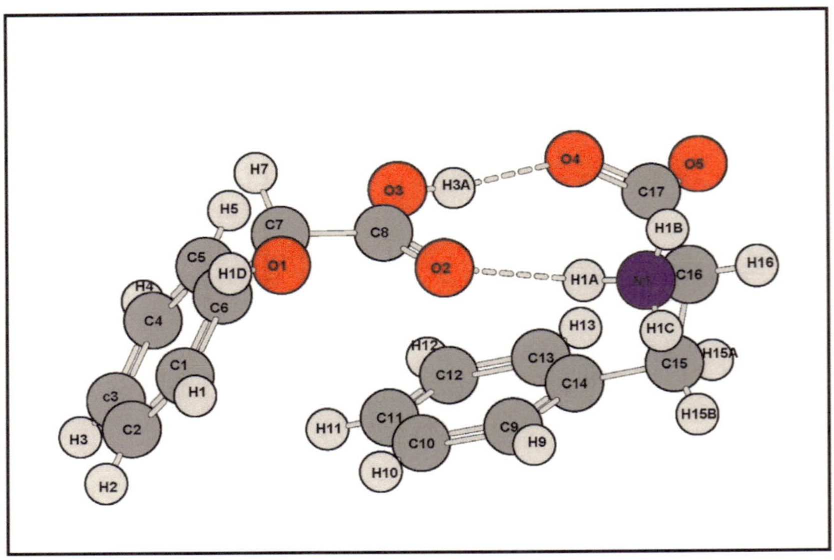

Figure 3.
Optimised labelled structure of RPASMA using MO6 DFT with 6-31++G(d,p).

4. Results and discussion

The compound RPASMA formed through donor-acceptor charge transfer mechanism during noncenterosymmetric monoclinic molecular crystal formation results in

Particulars	MADFT.out	LPADFT.out	RPASMADFT.out
File Type	.log	.log	.log
Calculation Type	FREQ	FREQ	FREQ
Calculation Method	RM06	RM06	RM06
Charge	0	0	0
Spin	Singlet	Singlet	Singlet
Solvation	None	None	None
E(RM06) Hartree	−535.04918	−554.4665	−1089.5313
RMS Gradient Norm Hartree/Bohr	2.14E-05	8.30E-06	1.59E-05
Imaginary Freq	0	0	0
Dipole Moment Debye	3.2937709	4.924759	7.0780304
Polarizability a.u.	100.668	117.08567	215.24667

Table 1.
Comparison of DFT details of MA, LPA and RPASMA using G09 MO6-31++G(d,p).

the variation of bond lengths, bond angles, dihedral angles and torsional angles than expected values. The hydrogen bonding interactions give distorted noncentero-symmetric structure during crystallisation. The XRD values and the DFT values are compared for the compounds MA, LPA and RPASMA and are reported in detail.

4.1 Bond length

The expected average C-C bond length is 1.54 Å, C-O bond length is 1.43 Å, C-N bond length is 1.43 Å, C-H bond length is 1.09 Å, C=O bond length is 1.23 Å, O-H bond length is 1.67 Å, N-H bond length is 2.1 Å. The comparative bond length study of the compounds show the C-C, C-H, C-O and O-H the equal XRD and DFT values shown in the **Table 2**. The hydrogen bonding existing in C-O, O-H, N-H, C-H shows considerable variation in values in RPASMA for the respective atoms of MA and LPA. N1-H9 bond length in LPA theoretical value is higher than experimental value due to the measurement in gaseous state and solid state and the zwitterionic nature of amino acid [14].

4.2 Bond angle (Å)

The intermolecular hydrogen bonding during the crystallisation of RPASMA give variations in bond angles due to sp^3 hybridisation and sp^2 hybridisation compounds because of their measurements in solid phase XRD and gaseous phase DFT are shown in the **Table 3** and theoretical values are due to the intramolecular hydrogen bonding and the respective measurements in the RPASMA are due to the internal strain and stress factors experienced during inter molecular hydrogen bonding. In the experimental and theoretical measurements of LPA show variation in bond angles due to the zwitter ionic nature of amino acid and the respective atoms in RPASMA show variation in the atoms involved in inter molecular hydrogen bonding which causes strain in the crystallisation [15].

		Experimental	Theoretical	Corresponding atoms in RPASMA	Experimental	Theoretical
Atoms in MA	C1-C2	1.507	1.519	C7-C8	1.521	1.525
	C1-O1	1.298	1.347	C8-O1	1.201	1.223
	C1-O2	1.197	1.203	C8-O2	1.298	1.308
	C2-H1	0.979	1.103	C7-H7	0.981	1.101
	C2-C3	1.505	1.514	C6-C7	1.516	1.514
	C2-O3	1.407	1.402	C7-O3	1.409	1.407
	C3-C4	1.366	1.397	C6-C5	1.381	1.395
	C3-C8	1.359	1.394	C6-C1	1.368	1.395
	C4-H5	0.93	1.088	C5-H5	0.93	1.091
	C4-C5	1.369	1.391	C4-C5	1.378	1.392
	C5-H4	0.929	1.087	C4-H4	0.931	1.089
	C5-C6	1.344	1.394	C4-C3	1.361	1.394
	C6-H3	0.931	1.087	C3-H3	0.931	1.088
	C6-C7	1.329	1.392	C3-C2	1.371	1.393
	C7-H2	0.931	1.087	C2-H2	0.931	1.088
	C7-C8	1.378	1.392	C2-C1	1.382	1.392
	C8-H6	0.929	1.089	C1-H1	0.931	1.088
	01-H7	0.819	0.971	O1-H1D	0.811	0.964
	O3-H8	0.819	0.964	O3-H3A	0.811	1.011
Atoms in LPA	C1-C2	1.519	1.533	C17-C16	1.532	1.557
	C1-O1	1.227	1.205	C17-O4	1.251	1.269
	C1-O2	1.251	1.336	C17-O5	1.229	1.227
	C2-H1	0.981	1.102	C16-H16	0.981	1.097
	C2-C3	1.519	1.539	C16-C15	1.527	1.523
	C2-N1	1.486	1.465	C16-N1	1.483	1.506
	C3-H2	0.969	1.099	C15-H15A	0.971	1.096
	C3-H3	0.971	1.098	C15-H15B	0.971	1.101
	C3-C4	1.499	1.507	C15-C14	1.502	1.505
	C4-C5	1.373	1.398	C14-C13	1.383	1.401
	C4-C9	1.376	1.397	C14-C9	1.381	1.395
	C5-H4	0.931	1.091	C13-H13	0.931	1.089
	C5-C6	1.369	1.392	C13-C12	1.381	1.391
	C6-H5	0.931	1.088	C12-H12	0.931	1.088
	C6-C7	1.368	1.392	C12-C11	1.361	1.393
	C7-H6	0.931	1.088	C11-H11	0.931	1.088
	C7-C8	1.359	1.392	C11-C10	1.356	1.391
	C8-H7	0.929	1.088	C10-H10	0.931	1.088

	Experimental	Theoretical	Corresponding atoms in RPASMA	Experimental	Theoretical
C8-C9	1.375	1.392	C10-C9	1.377	1.392
C9-H8	0.931	1.088	C9-H9	0.931	1.092
N1-H9	0.895	1.959	N1-H1A	0.891	1.052
N1-H10	1.006	1.014	N1-H1B	0.941	1.025
N1-H11	0.884	1.018	N1-H1C	0.881	1.021

Table 2.
Bond length (Å).

		Experimental	Theoretical	Corresponding atoms in RPASMA	Experimental	Theoretical
Atoms in MA	C2-C1-O1	111.576	110.029	C7-C8-O1	104.651	106.701
	C2-C1-O2	124.451	126.239	C7-C8-O2	122.801	122.141
	O1-C1-O2	123.961	123.677	O1-C8-O2	124.361	126.171
	C1-C2-H1	108.204	107.312	C8-C7-H7	109.101	106.451
	C1-C2-C3	112.203	108.162	C8-C7-C6	111.431	110.231
	C1-C2-O3	108.211	107.248	C8-C7-O3	112.801	111.681
	H1-C2-C3	108.221	108.863	H7-C7-C6	109.101	108.771
	H1-C2-O3	108.192	111.824	H7-C7-O3	109.045	111.395
	C3-C2-O3	111.681	113.194	C6-C7-O3	113.482	113.073
	C2-C3-C4	120.832	119.431	C7-C6-C5	119.401	120.101
	C2-H1-C8	121.411	120.819	C7-H7-C1	121.601	120.281
	C4-C3-C8	117.744	119.746	C5-C6-C1	119.001	119.611
	C3-C4-H5	119.749	119.095	C6-C5-H5	119.901	120.261
	C3-C4-C5	120.427	119.984	C6-C5-C4	120.301	120.281
	H5-C4-C5	119.824	120.921	H5-C4-C5	119.901	120.261
	C4-C5-H4	119.497	119.814	C4-C5-H4	119.701	120.141
	C4-C5-C6	120.937	120.141	C4-C5-C3	120.501	120.001
	H4-C5-C6	119.566	120.045	H4-C5-C3	119.421	120.103
	C4-C5-H3	120.252	120.034	C5-C3-H3	109.053	108.774
	C4-C5-C7	119.411	119.961	C5-C3-C2	111.424	108.774
	H3-C5-C7	120.336	120.005	H3-C3-C2	120.301	120.121
	C5-C7-H2	119.635	120.178	C3-C2-H2	119.801	120.041
	C5-C7-C8	120.662	119.957	C3-C2-C1	120.501	120.201
	H2-C7-C8	119.704	119.864	H2-C2-C1	119.801	119.761
	C3-C8-C7	120.802	120.209	C6-C1-C2	120.201	120.071

		Experimental	Theoretical	Corresponding atoms in RPASMA	Experimental	Theoretical
	C3-C8-H6	119.631	119.517	C6-C1-H1	119.901	119.101
	C7-C8-H6	119.568	120.274	C2-C1-H1	119.901	120.841
	C1-O1-H7	109.441	107.042	C8-O1-H1D	114.798	108.093
	C2-O3-H8	109.459	108.257	C7-O3-H3A	107.024	112.549
Atoms in LPA	C2-C1-O1	118.488	123.816	C16-C17-O4	118.401	113.841
	C2-C1-O2	116.555	113.748	C16-C17-O5	117.201	115.261
	O1-C1-O2	124.906	122.421	O4-C17-O5	124.301	130.901
	C1-C2-H1	109.025	104.789	C17-C16-H16	107.701	108.391
	C1-C2-C3	111.507	112.325	C17-C16-C15	111.201	113.161
	C1-C2-N1	109.834	108.267	C17C16-N1	110.571	107.111
	H1-C2-C3	109.055	108.249	H16-C16-C15	107.701	110.161
	H1-C2-N1	109.029	106.434	H16-C16-N1	107.701	106.891
	C3-C2-N1	108.351	116.027	C15-C16-N1	111.701	110.891
	C2-C3-H2	109.061	107.115	C16-C15-H15A	108.401	105.381
	C2-C3-H3	109.016	109.336	C16-C15-H15B	108.401	110.091
	C2-C3-C4	113.019	113.935	C16-C15-C14	115.381	112.841
	H2-C3-H3	107.729	107.264	H15A-C15-H15B	107.501	107.301
	H2-C3-C4	108.927	108.417	H15A-C15-C14	108.401	111.031
	H3-C3-C4	108.953	110.504	H15B-C15-C14	108.401	109.991
	C3-C4-C5	120.524	119.383	C15-C14-C13	120.801	120.521
	C3-C4-C9	121.631	121.944	C15-C14-C9	121.601	120.511
	C5-C4-C9	117.834	118.666	C13-C14-C9	117.601	118.931
	C4-C5-H4	119.346	119.315	C14-C13-H13	119.501	119.321
	C4-C5-C6	121.301	120.995	C14-C13-C12	120.901	120.251
	H4-C5-C6	119.354	119.687	H13-C13-C12	119.501	120.421
	C5-C6-H5	120.147	119.944	C13-C12-H12	119.901	119.711
	C5-C6-C7	119.766	119.872	C13-C12-C11	120.201	120.371
	H5-C6-C7	120.087	120.182	H12-C12-C11	119.901	119.921
	C6-C7-H6	119.976	120.178	C12-C11-H11	120.201	120.061
	C6-C7-C8	120.087	119.598	C12-C11-C10	119.701	119.771
	H6-C7-C8	119.936	120.224	H11-C11-C10	120.201	120.161
	C7-C8-H7	120.041	120.024	C11-C10-H10	119.701	120.111
	C7-C8-C9	119.768	120.411	C11-C10-C9	120.501	119.871
	H7-C8-C9	120.191	119.565	H10-C10-C9	119.701	120.021
	C4-C9-C8	121.223	120.458	C14-C9-C10	121.101	120.801
	C4-C9-H8	119.384	119.035	C14-C9-H9	119.501	119.391

	Experimental	Theoretical	Corresponding atoms in RPASMA	Experimental	Theoretical
C8-C9-H8	119.393	120.404	C10-C9-H9	119.501	119.811
C2-N1-H9	112.675	82.203	C16-N1-H1A	113.501	112.231
C2-N1-H10	110.662	111.593	C16-N1-H1B	107.801	107.301
C2-N1-H11	109.646	111.219	C16-N1-H1C	112.701	113.631
H9-N1-H10	105.982	133.369	H1A-N1-H1B	109.001	102.361
H9-N1-H11	106.483	107.774	H1A-N1-H1C	105.001	109.561
H10-N1-H11	111.294	107.603	H1B-N1-H1C	108.001	111.161

Table 3.
Bond angle (Å).

4.3 Dihedral angle

Dihedral angle value increases in the RPASMA compared to MA and LPA, it shows decrease in value where the hydrogen bonding takes place. The negative values in the case of molecular crystal is due to the formation non-centerosymmetric organic salt having delocalisation of charge. The steric hindrance give distorted ring structure is shown in the **Table 4**. The decrease in torsional values of salt compared to the MA and LPA shows the high symmetry attained during crystallisation due to hydrogen bonding interactions. The charge transfer mechanism during crystallisation result in negative values [16].

		Experimental	Theoretical	Corresponding atoms in RPASMA	Experimental	Theoretical
Atoms in MA	O1-C1-C2-C3	59.381	81.265	O1-C8-C7-C6	69.900	94.817
	O1-C1-C2-O3	−176.954	−156.333	O1-C8-C7-O3	−167.100	−167.106
	O2-C1-C2-C3	−121.821	−96.124	O2-C8-C7-C6	−110.800	−110.852
	O2-C1-C2-O3	1.844	26.279	O2-C8-C7-O3	12.200	38.300
	C1-C2-C3-C4	69.561	73.525	C8-C7-C6-C5	−70.900	−70.832
	O3-C2-C3-C4	−52.141	−45.164	O3-C7-C6-C5	171.300	150.782
	O2-C2-C3-C8	126.482	135.629	O3-C7-C6-C1	−6.900	−28.943
	C8-C3-C4-C5	−0.900	−0.560	C1-C6-C4-C5	−1.600	0.480
	C2-C3-C8-C7	−177.848	−179.502	C7-C6-C1-C2	179.200	178.505
	C3-C4-C5-C6	1.185	0.425	C6-C4-C5-C3	1.300	0.561
	C4-C5-C6-C7	−1.347	−0.024	C4-C5-C3-C2	−0.300	−0.862
	C6-C5-C8-C3	−1.014	0.103	C3-C2-C1-C6	0.100	0.924
Atoms in LPA	O1-C1-C2-C3	−91.018	−24.481	O5-C17-C16-C15	−75.300	−54.612
	O1-C1-C2-N1	148.858	−153.905	O5-C17-C16-N1	160.000	−177.121
	O2-C1-C2-C3	86.524	156.910	O4-C17-C16-C15	103.200	125.749
	O2-C1-C2-N1	−33.601	27.485	O4-C17-C16-N1	−21.500	3.240

	Experimental	Theoretical	Corresponding atoms in RPASMA	Experimental	Theoretical
C1-C2-C3-C4	70.644	88.972	C17-C16-C15-C14	−56.400	−63.318
N1-C2-C3-C4	−168.363	−145.745	N1-C16-C15-C14	67.600	57.061
C2-C3-C4-C5	81.910	90.273	C16-C15-C14-C9	−84.500	−102.871
C2-C3-C4-C9	−99.128	−88.697	C16-C15-C14-C13	94.900	74.611
C9-C4-C5-C6	1.012	0.228	C13-C14-C9-C10	−0.300	−0.811
C3-C4-C9-C8	−179.174	178.940	C15-C14-C13-C12	−179.600	−176.669
C5-C4-C9-C8	−0.185	−0.037	C9-C14-C13-C12	−0.200	0.853
C4-C5-C6-C7	−1.703	−0.263	C14-C9-C10-C11	0.500	0.359
C5-C6-C7-C8	1.546	0.105	C9-C10-C11-C12	−0.200	0.060
C6-C7-C8-C9	−0.736	0.083	C10-C11-C12-C13	−0.300	−0.014
C7-C8-C9-C4	0.055	−0.117	C11-C12-C13-C14	0.500	−0.449

Table 4.
Dihedral angle (Å).

4.4 Bond alteration coefficient

The Bond Alteration Coefficient (BAC) of the compounds show the solid state experimental measurements of MA and RPASMA are almost equal but the theoretical gaseous phase measurements are varying due to van der Waals forces of attraction. The BAC of LPA and RPASMA are showing variations in both solid and gaseous phase measurements indicate the zwitter ionic nature of LPA leads to charge transfer mechanism as shown in the **Table 5**.

		Exp	BAC	DFT	BAC	RPASMA	Exp	BAC	DFT	BAC
MA	C1-C2	1.507		1.519		C17-C16	1.521		1.525	
	C2-C3	1.505	0.002	1.514	0.005	C16-C15	1.516	0.005	1.514	0.009
	C3-C4	1.366	0.139	1.397	0.117	C15-C14	1.381	0.135	1.395	0.119
	C4-C5	1.369	−0.003	1.391	0.006	C14-C13	1.378	0.003	1.392	0.003
	C5-C6	1.344	0.025	1.394	−0.003	C13-C12	1.361	0.017	1.394	0.002
			0.166		0.128			0.16		0.133
LPA	C4-C5	1.376		1.397		C13-C14	1.383		1.401	
	C5-C6	1.369	0.007	1.392	0.005	C12-C13	1.381	0.002	1.391	0.01
	C6-C7	1.368	0.001	1.392	0	C11-C12	1.361	0.02	1.393	−0.002
	C7-C8	1.359	0.009	1.392	0	C10-C11	1.356	0.005	1.391	0.002
	C8-C9	1.375	0.016	1.392	0	C9-C10	1.377	0.021	1.392	−0.001
			0.033		0.005			0.048		0.012

Table 5.
Bond alteration coefficient analysis of the title compounds.

4.5 Hydrogen bonding

The hydrogen bonding in XRD values and DFT values confirm the charge transfer mechanism happening between donor- acceptor interactions resulting in the formation non centerosymmetric crystallisation [17] as shown in the **Table 6**.

4.6 Mulliken atomic charges for MA, LPA and RPASMA

Mulliken atomic charges of the compounds show large positive charges of the acceptor hydrogen atoms and high negative charges of the donor O atom and N atoms respectively. MA and RPASMA show decrease in charge values respectively but the carbon atoms close to hydrogen bonding show increase in the atomic charge value. H1A of RPASMA connected to nitrogen atom show higher positive charge compared to LPA because of charge transfer interactions during hydrogen bonding [18]. The Mulliken atomic charges show the presence of charge transfer mechanism in the crystal formation as shown in the **Table 7**.

D – H….. A	d(D-H)		d(H…..A)		d(D…..A)		∠DHA	
	Expt.	Theo.	Expt.	Theo.	Expt.	Theo.	Expt.	Theo.
C(15) – ….. O(5)	0.970	1.096	2.877	2.597	3.072	2.928	92.27	96.32
O(3) – H(3A) ….. O(4)	0.812	1.011	1.807	1.606	2.605	2.607	167.15	116.70
N(1) – H(1A) ….. O(2)	0.886	1.052	2.057	1.747	2.918	2.744	163.59	156.50
N(1) – H(1B) ….. O(4)	0.940	1.025	2.819	2.213	2.704	2.529	73.32	95.68

Table 6.
Hydrogen bonding in RPASMA.

MA ATOM	Charge	Corresponding atoms in RPASMA	Charge	LPA ATOM	Charge	RPASMA ATOM	Charge
C1	0.246	C8	−0.267	C1	0.405	C17	0.108
C2	0.13	C7	0.744	C2	−0.518	C16	−0.176
C3	−0.648	C6	−1.259	C3	0.079	C15	0.055
C4	0.699	C5	0.094	C4	−0.082	C14	−0.544
C5	−0.114	C4	−0.219	C5	−0.107	C13	0.214
C6	0.045	C3	−0.001	C6	−0.112	C12	−0.004
C7	−0.246	C2	0.006	C7	0.002	C11	−0.289
C8	−0.419	C1	0.19	C8	−0.297	C10	−0.034
H1	0.137	H7	0.132	C9	−0.016	C9	0.081
H2	0.144	H2	0.132	H1	0.199	H16	0.171
H3	0.112	H3	0.134	H2	0.147	H15A	0.217
H4	0.107	H4	0.129	H3	0.164	H15B	0.107
H5	0.109	H5	0.128	H4	0.075	H13	0.164
H6	0.094	H1	0.137	H5	0.1	H12	0.137

MA ATOM	Charge	Corresponding atoms in RPASMA	Charge	LPA ATOM	Charge	RPASMA ATOM	Charge
H7	0.4	H1D	0.385	H6	0.098	H11	0.047
H8	0.388	H3A	0.556	H7	0.104	H10	0.131
O1	−0.416	O1	−0.407	H8	0.15	H9	0.081
O2	−0.375	O2	−0.339	H9	0.416	H1A	0.506
O3	−0.393	O3	−0.333	H10	0.335	H1B	0.403
				H11	0.332	H1C	0.395
				O1	−0.469	O4	−0.641
				O2	−0.395	O5	−0.429
				N1	−0.61	N1	−0.642
Total charge	0		−0.058		0		0.058

Table 7.
Mulliken atomic charges for MA, LPA and RPASMA.

4.7 Proton NMR

The non centerosymmetric structure of the RPASMA show the chemical shifts for the various type of protons as expected for both shielding effect and deshielding effect as shown in the **Table 8**. The OH proton present in the H-C-OH environment

Hydrogen atom number	Position in RPASMA	Experiment (ppm)	Theoretical (ppm)
H1	MA aromatic ring	7–8	9.65
H2	MA aromatic ring	7–8	7.98
H3	MA aromatic ring	7–8	7.47
H4	MA aromatic ring	7–8	7.57
H5	MA aromatic ring	7–8	9.5
H1D	H in H-C-OH in MA	5	6.13
H7	H in OH in H-C-OH in MA	3.7	2.9
H3A	H in COOH in MA	7–8	9.7
H9	LPA aromatic ring	7–8	7.64
H10	LPA aromatic ring	7–8	8.26
H11	LPA aromatic ring	7–8	6.14
H12	LPA aromatic ring	7–8	7.99
H13	LPA aromatic ring	7–8	9.7
H15A	CH_2 in LPA	7–8	4.32
H15B	CH_2 in LPA	3.7	4.31
H16	H in C connected to N in LPA	3.8	4.1

Hydrogen atom number	Position in RPASMA	Experiment (ppm)	Theoretical (ppm)
H1A	H in $NH3^+$ in LPA connected to MA	5.1	7.63
H1B	H in $NH3^+$ in LPA	5.1	7.5
H1C	H in $NH3^+$ in LPA	5.1	7.2

Table 8.
Proton NMR.

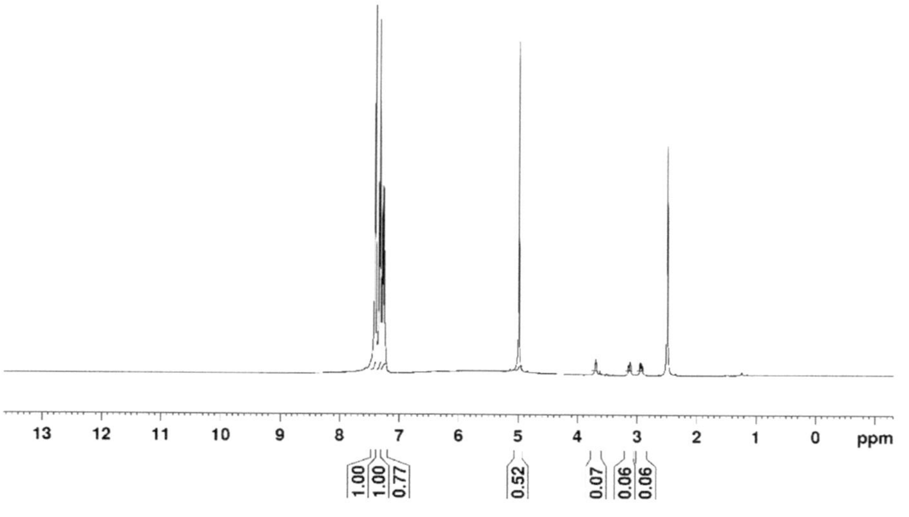

Figure 4.
Proton NMR of RPASMA.

and the COOH environment close to each other. The molecular strain in the molecule due to hydrogen bonding leads to the values for both experimental and theoretical values in the measurements of XRD and DFT respectively as shown in the **Figures 4** and 5.

4.8 Carbon NMR

Carbon NMR calculated to TMS show similar values for both experimental and theoretical values of chemical shifts, **Table 9**. The experimental values are less compared to the theoretical values because of the method of estimation in the solid state and gaseous state respectively. Aromatic carbon shifts show the increase in theoretical values, the functional group carbon atoms show almost same or decrease in theoretical values may be due to the charge delocalisation in solid and gaseous phases respectively.

4.9 FTIR

The non linear aromatic molecules obey the 3N-6 rule for fundamental modes of vibrations, (N is the total number of atoms in the molecule). The number of atoms in

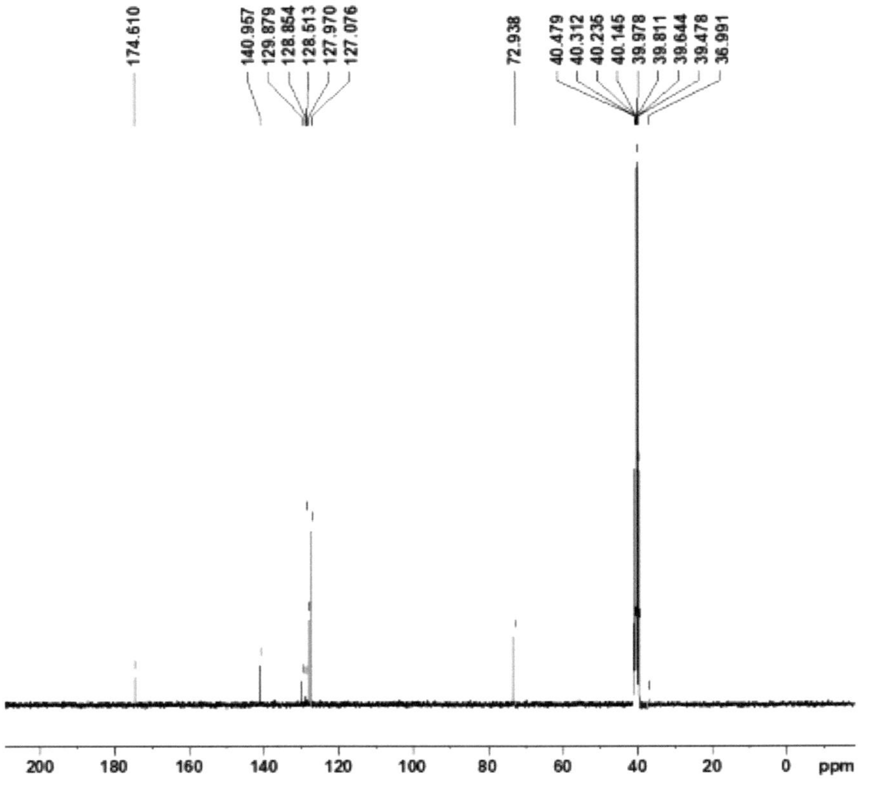

Figure 5.
Carbon NMR of RPASMA.

Carbon atom number	Position in RPASMA	Experiment (ppm)	Theoretical (ppm)
C1	MA aromatic ring	127.08	143.97
C2	MA aromatic ring	127.97	137.71
C3	MA aromatic ring	128.51	145.51
C4	MA aromatic ring	128.85	141.83
C5	MA aromatic ring	129.88	139.73
C6	MA aromatic ring connected to C7	129.88	164.08
C7	H-C-OH in MA	72.93	73.28
C8	COOH in MA	174.51	157.17
C9	LPA aromatic ring	127.08	122.26
C10	LPA aromatic ring	127.97	136.26
C11	LPA aromatic ring	128.51	144.24
C12	LPA aromatic ring	128.85	142.21
C13	LPA aromatic ring	129.88	145.84
C14	LPA aromatic ring	129.88	154.76

Carbon atom number	Position in RPASMA	Experiment (ppm)	Theoretical (ppm)
C15	LPA -CH_2	129.88	146.33
C16	LPA -C-N	36.99	60.94
C17	COOH in LPA	140.96	120.39

Table 9.
Comparative carbon NMR data of RPASMA.

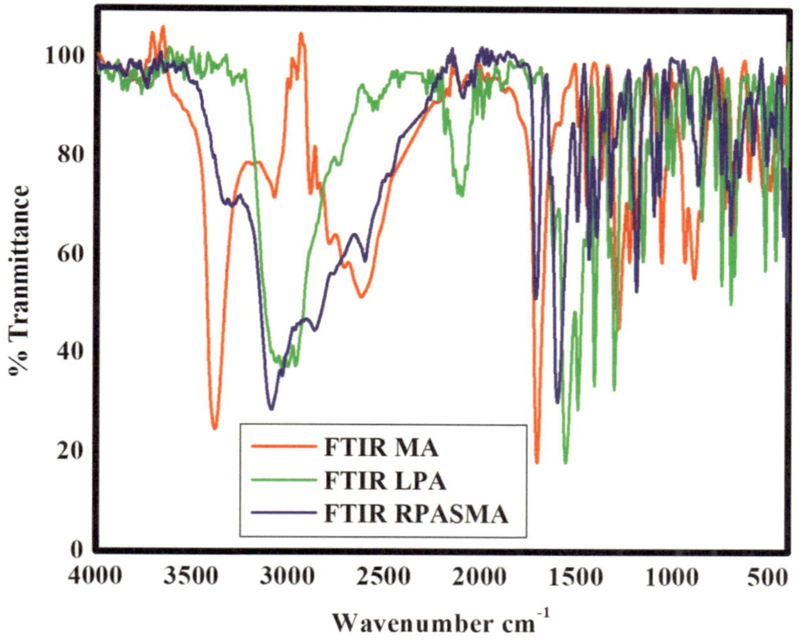

Figure 6.
Comparison of FTIR of the compounds.

MA are 19 and the fundamental modes of vibrations expected are 51, also confirmed by DFT study. The presence of total number of 23 atoms in LPA show 63 fundamental modes of vibrations by FTIR, which has been confirmed by DFT study. The RPASMA molecule with 42 atoms which shows 120 modes of fundamental vibrations confirmed by DFT study. The comparative FTIR PLOT of the title compounds is shown in the **Figure 6**.

4.10 FT Raman

The polarisation effect leading to dipole change is measured using FT Raman. The charge transfer mechanism in the RPASMA is confirmed by the FT Raman study and is shown in the **Figure 7** and the respective values are given in the **Table 10**.

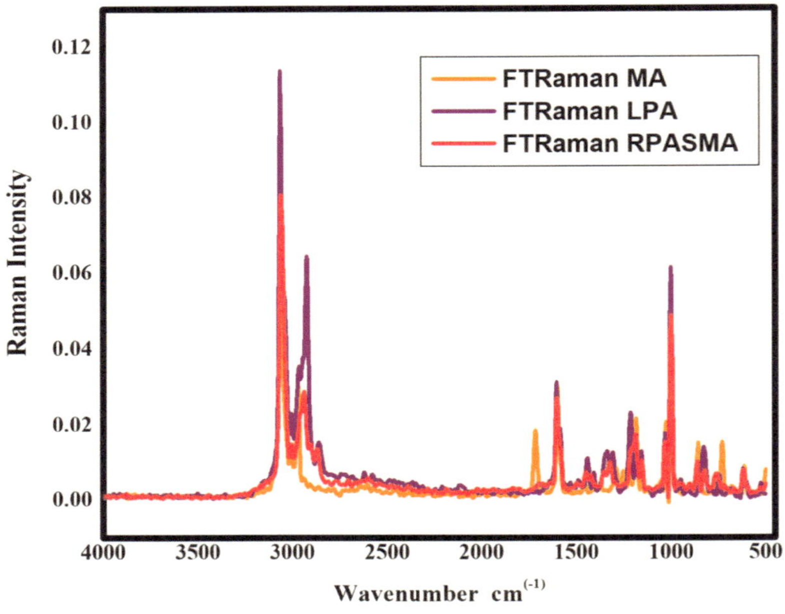

Figure 7.
Comparison of the compounds FT Raman study.

Wavenumber (cm^{-1})	MA	LPA	RPASMA
3060	0.092	0.110	0.081
2939	0.033	0.065	0.029
1604	0.033	0.032	0.028
1189	0.022	0.012	0.017
1032	0.017	0.013	0.016
1000	0.055	0.059	0.048
861	0.016	0.014	0.010
822	0.002	0.013	0.009
757	0.016	0.013	0.008
617	0.010	0.008	0.008

Table 10.
Comparison of Raman intensities of MA, LPA and RPASMA.

5. Thermochemical properties

The thermochemical properties of the compounds indicate the spontaneity of the reaction and it is accompanied by the decrease in free energy. As the polarisation increases due to charge transfer mechanism, donor – acceptor interactions, the dipole moment increases in the compound after crystallisation as bimolecular single crystal (**Table 11**).

The comparison of HOMO – LUMO energy details **Figures 8** and **9** show the low energy gap in RPASMA formation from the donor – acceptor interactions of MA and

Parameter	MA	LPA	RPASMA
Total Energy (E) Hartree	−535.04918	−554.4665	−1089.5313
Zero point Vibrational Energy Kcal/Mole	93.1878	119.2589	213.9199
Dipole moment (Debye)	3.2938	4.9247	7.0780
Total heat capacity Cal/Mole-Kelvin	36.5180	41.4870	81.5680
Total Entropy Cal/Mole-Kelvin	97.6190	104.5540	151.3150
Total thermal Energy Kcal/Mol	99.2460	126.0270	227.1680

Table 11.
Comparison of thermochemical properties of MA, LPA and RPASMA.

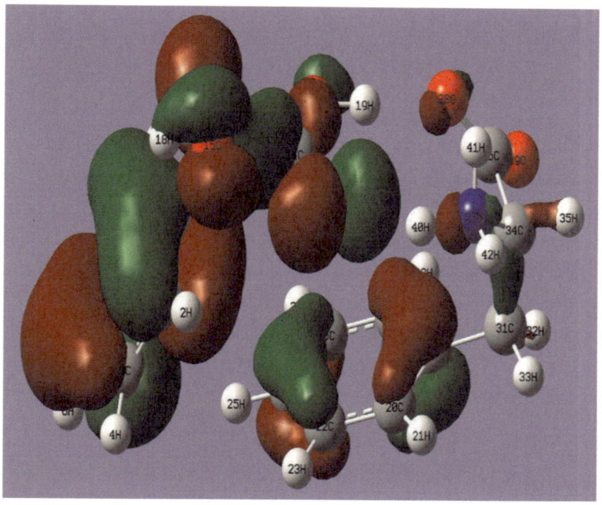

Figure 8.
RPASMA HOMO PLOT.

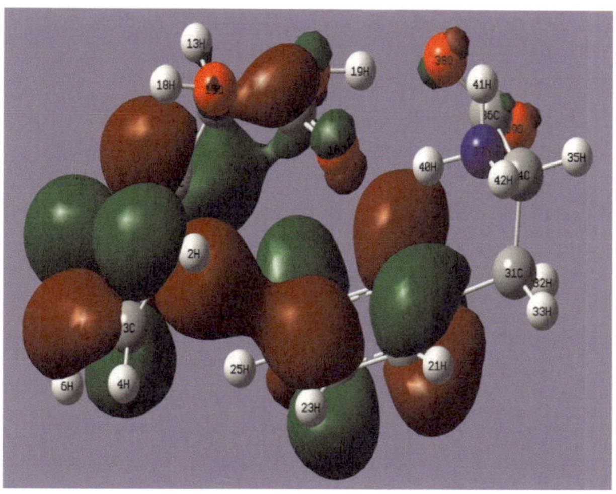

Figure 9.
RPASMA LUMO PLOT.

Parameter	MA	LPA	RPASMA
HOMO (Orbital)	40	44	84
HOMO (Energy)	−0.27163	−0.25791	−0.25269
LUMO (Orbital)	41	45	85
LUMO (Energy)	−0.04288	−0.05366	−0.06529
Energy Gap	−0.22875	−0.20425	−0.1874

Table 12.
Comparison of HOMO – LUMO energies of MA, LPA and RPASMA.

LPA, which is resulting in the formation of less stable and highly reactive compound as shown in the **Table 12** [19, 20].

6. Conclusion

The donor – acceptor interactions in the MA and LPA, give charge transfer mechanism in monoclinic RPASMA. The comparative study of the compounds show the parameters of solid state XRD measurements and the gaseous state DFT measurements are comparable. This study also confirms the product having inter and intra molecular hydrogen bonding, polarisation and van der Waals forces of attractions. The C2 space group results in noncenterosymmetric crystal structure. The FTIR vibrational study confirms the XRD and DFT parameters. The low energy gap of the RPASMA results in highly reactive nature and possibilities of compound in biological activity.

Conflicts of interest

None.

Author details

Subramanian Usha[1*] and Charles Kanakam Christopher[2]

1 Department of Chemistry, Sri Sairam Engineering College, Chennai, India

2 Department of Chemistry, Presidency College, Chennai, India

*Address all correspondence to: usha.che@sairam.edu.in

References

[1] Usha S, Charles KC. Synthesis, characterisation of novel NLO material: Bis-L-phenyl alanine mandelate. Chemical Science Transactions. 2016; **5**(1):179-186

[2] Usha S, Charles KC. A novel method of diastereomeric resolution of R-phenyl alanine-S-mandelate from the racemic mixture. Chemical Science Transactions. 2016;**5**(3):727-733

[3] Usha S, Charles. Kanakam Christopher. Theoretical DFT study of R-phenyl alanine-S-mandelate. Materials Today: Proceedings. 2019;**16**:1137-1145

[4] Jacob PL, Heinz-Jurgen B. Synthesis of polyhydroxylated aromatic mandelic acid amides and their antioxidative potential. Tetrahedron. 2001;**57**:1277-1282

[5] Anna M, Ewa C, Marian Z. Estimation of the antioxidative properties of amino acids – An electrochemical approach. International Journal of Electrochemical Science. 2014;**9**:7904-7915

[6] Rebecca KH, Graham J, Tizzard Terence L, Threlfall Amy L, Simon JC, Colin CS, et al. Are the crystal structures of enantiopure and racemic mandelic acids determined by kinetics or thermodynamics? Journal of American Chemical Society. 2015;**137**:11095-11104

[7] Nabil E, Walid A, Ahmed AB, Younes A, Habib F. Vibrational spectroscopic study, charge transfer interaction and nonlinear optical properties of L-asparaginium picrate: A density functional theoretical approach, SAA Molecular and Biomolecular Spectroscopy. 2014;**128**:781-789

[8] Usha S, Charles KC. Antioxidant activity of phenyl alanine mandelates by chemical and electrochemical methods. Global Journal of Medical Research (B). 2016;**16**(1):1-6

[9] Peter IN. Competing intramolecular *vs.* intermolecular hydrogen bonds in solution. International Journal of Molecular Science. 2014;**15**:19562-19633

[10] Savita SK, Ashish RN, Näther C, Wolfgang B, Bikshandarkoil RS. Synthesis, crystal structure and photochemistry of Hexakis(butan-1-aminium) heptamolybdate(VI) tetrahydrate. Journal of Chemical Sciences. 2016;**128**(11):1737-1744

[11] Evecen M, Tanak H. Quantum chemical studies on the molecular structure, spectroscopic and electronic properties of (6-Methoxy-2-oxo-2H-chromen-4-yl)-methyl pyrrolidine-1-carbodithioate. Material Science Poland. 2016;**34**(4):886-904

[12] Hyunwoo L, Joshua HB, Robert WF, John FS. High-accuracy estimates for the Vinylidene-acetylene isomerization energy and the ground state rotational constants of :C=CH2. Journal of Physical Chemistry A. 2013;**117**:11679-11683

[13] Rohan JK, Jegadesan S, Andrew BH. Enhancement of efficiency in organic photovoltaic devices containing self-complementary hydrogen-bonding domains. Beilstein Journal of Organic Chemistry. 2013;**9**:1102-1110

[14] Priyanka PK, Prashant KB, Babasaheb RS, Dipak SD. Synthesis of D-D-A-type small organic molecules with enlarged linker system towards organic solar cells and effect of co-adsorbents on cell performance. New Journal of Chemistry. 2016;**40**:634-640

[15] Peter FL, Barry CS. Single bond lengths of organic molecules in the solid

state. Global Journal of Science Frontier. 2016;**16**(3):55-60

[16] Sreenivasa S, Palakshamurthy BS, Pampa KJ, Lokanath NK, Suchetane PA. 2-Chloro-N-(4- methoxybenzoyl) benzenesulfonamide. Acta Crystal E. 2014;**70**:199

[17] Jessy E, Sithambaresan M, Prathapachandra Kurup MR. N0-[(E)-Furan-2-ylmethylidene]pyridine-3-carbohydrazide. Acta Crystal E. 2011;**67**: 3267

[18] Motohiro N, Yoji U, Kazumasa H, Sei T, Hiroko S. CH/p hydrogen bonds in organic and organometallic chemistry. Crystal Engineering Communication. 2009;**11**:1757-1788

[19] Mathammal R, Jayamani N, Geetha N. Molecular structure, NMR, HOMO, LUMO and vibrational analysis of O-anisic acid and anisic acid based on DFT calculations, Journal of Spectroscopy. 2013;**2013**:1-18

[20] Monika P, Renata S, Grzegorz S, Marzena M, Włodzimierz L. Spectroscopic characterization and antioxidant properties of mandelic acid and its derivatives in a theoretical and experimental approach. Materials. 2022; **15**:5413-5433